BOAT CARPENTRY

REPAIRS · ALTERATIONS · CONSTRUCTION

SECOND EDITION

HERVEY GARRETT SMITH

 VAN NOSTRAND REINHOLD COMPANY

NEW YORK   CINCINNATI   ATLANTA   DALLAS   SAN FRANCISCO
LONDON   TORONTO   MELBOURNE

Published by VAN NOSTRAND REINHOLD COMPANY
135 West 50th Street, New York, N.Y. 10020

VAN NOSTRAND REINHOLD LIMITED
1410 Birchmount Road
Scarborough, Ontario M1P 2E7, Canada

VAN NOSTRAND REINHOLD AUSTRALIA PTY. LTD.
17 Queen Street
Mitcham, Victoria 3132, Australia

VAN NOSTRAND REINHOLD COMPANY LIMITED
Molly Millars Lane
Wokingham, Berkshire, England

16 15

To
The Memory of
My Father
Whose Precepts Furnished the Inspiration
for This Book

Preface

Although the preface appears at the beginning of a book, in this instance it was the last thing written. In looking over my efforts, I discover that very little of the information herein has been obtained from books or other outside sources. For the most part, it consists of things I have learned from experience as an amateur, in the building, maintenance and repair of boats.

My interest in the techniques of boat carpentry stems from the teachings of my father, a country builder and craftsman of the old school, whose methods of instruction were most effective, though not always painless. When I was a boy in my early teens, he utterly destroyed my first attempt at building a boat with one fell sweep of an axe, more in pity than in anger. Yet before my tears were scarcely dry he was building me a new one—just to show me how it *should have* been done. From him I learned how to use tools, and how to work with wood, and to have a sincere respect for both. He was free with help and criticism alike, and I was a very fortunate lad.

I know full well the trials and tribulations that beset the amateur. Along with his limited experience, he invariably lacks proper tools and equipment, and seldom has adequate space in which to work. The success with which he overcomes these deficiencies depends entirely on the limits of his ingenuity and his ability to improvise. All too often he has no one to turn to for qualified advice or help.

It is for just such an amateur that this book was written. It is an attempt to provide the answers to some of the problems I and many other amateurs have had to solve. Most of the recommendations and opinions are out of my own experience.

Boat construction is a highly specialized skill, to which house-building methods contribute little or nothing. If an amateur is to acquire this skill, he must have a clear understanding of the fundamental techniques that make it so specialized. This book does not provide all the answers, nor will it serve as a substitute for practical experience. But whether you undertake the repair, alteration or building of a boat, these are the elementary things you must know if you are to work intelligently.

It must be remembered that professional boat builders may and often do use methods which differ in some respects from those I have described in this book. This is largely due to the fact that they have equipment and facilities not ordinarily available to the amateur. In the light of this, I have endeavored to confine my suggestions to procedures which are entirely practical for the average amateur.

HERVEY GARRETT SMITH

July 15, 1965
Sayville, N. Y.

Acknowledgements

I wish to express my sincere appreciation to the International Nickel Company, The Borden Company, United States Plywood Corporation, and to Ichabod T. Williams & Sons, for their generous help and co-operation in assembling much of the technical material contained in this book.

Contents

I

Tools

It is an old truism that a conscientious workman takes pride in his tools, whatever his trade or profession. In fact, you can generally judge the worth of a man by the condition of his tools. Good tools are conducive to craftsmanship, while incompetence and carelessness go hand in hand with the rusty saw and the dull chisel. These truths apply to the amateur just as much as they do to the professional.

Born into a woodworking family, since childhood I have had access to practically every woodworking tool you can name, and at an early age I was taught, sometimes painfully, to regard them with respect. Therefore it is pardonable, I think, that I have always looked upon amateurs' tools with a rather jaundiced eye. I have always had the feeling that most amateurs don't know a good tool when they see it, nor a dull tool when they use it.

All this is rapidly becoming a thing of the past, thank the Lord, for with the advent of the "do-it-yourself" era, amateurs are beginning to take these things seriously. They are learning to judge tools, to care for them, and to use them in a professional manner. From haunting the bargain counters they have progressed to shopping for quality with a critical eye. In short, they have learned that you can't be a craftsman unless you have craftsmen's tools, and I think this is a wonderful thing.

So in this discussion of boat woodworking tools for the amateur, we can assume that we are both thinking of the same thing—tools of professional quality. What I am most interested in right here is a complete list of *useful* tools, tools that are needed in the construction, alteration or repair of boats, from the standpoint of the amateur and his limitations. Not all of these tools are indispensable— some are required only seldom, but all of them make the job easier.

If you were to acquire all of these tools at once, the cost would probably seem all out of proportion to the immediate job at hand, but the average man generally assembles his collection of tools piecemeal, anyway. The point to remember is that when your budget is limited, two chisels of high-quality tool steel are better than a set of eight poor ones that won't hold an edge.

Hammers

Hammers come in a variety of weights and styles, so they must be chosen for the work they are to do. It is desirable to have two carpenter's hammers—a size #2, with a 13 oz. head, for general use, and a size #1½, with a 16 oz. head for heavy driving. The heads should be drop-forged steel, tempered and hardened, with *curved* claws —not straight—and hickory handles.

A ball-pein hammer is needed, not only for riveting, but for heading over drifts. For this work the lighter ones are preferred, since it requires light taps, rather than heavy blows. The ½ or ¾ pound head is about right.

A light weight maul is needed for drifts. Drifts must be driven with the minimum number of heavy blows, and a hammer is too light for the job, for it will cause the drift to freeze before it's fully driven.

A wooden mallet is a must—for use with chisels. Size is not important, but weight is. Mallets are made of a variety of woods, and the denser, harder and heavier the wood the better. The best wood is lignum-vitae. This is a tool you can make yourself. All you need is a chunk of hickory, locust or other dense and heavy wood about 3 in. in diameter and 5 in. long, with an 8 in. handle. If you are going to do any heavy work with a chisel, *don't* strike it with a hammer—use the mallet; it is easier on the chisel and won't bounce off in a glancing blow as the hammer will.

Saws

Since so much of your work is done with hand saws, they should be chosen carefully, sharpened and set correctly, and maintained in top condition at all times. I consider two crosscut and one rip saw a minimum requirement. All are generally 26 in., but for interior cabinetwork of a light nature, your fine-toothed crosscut might be a 24 in. one. Crosscut saws for rough work should have 8 or 9 teeth to the inch, but for finished work, where clean-edged cuts are necessary for tight joints, 11 points to the inch are a must. Coarse toothed saws tear the wood as they cut, particularly in coarse-grained wood, such as fir plywood.

The rip saw has 5½ teeth per inch. If you have a choice, pick the rip saw with the greatest width or depth of blade. This makes it easier to saw in a straight line, with the least tendency to run off at a tangent.

A compass saw is needed frequently, particularly in repair work. They are made with two interchangeable blades—a 12 in. compass and a 10 in. keyhole. They are generally coarse toothed, and hence make a very rough cut, but they are indispensable in tight spots where you can't use a hand saw.

All hand saws should be hollow ground. This helps to prevent binding when making rather long cuts.

Good saws are supposed to be properly sharpened and set before they leave the factory, and I suppose most of them are. But I have noticed that skilled craftsmen of the old school will often file and set a new saw very soon after they buy it. Saw sharpening is an art, and it takes a lot of skill and experience to do it well. I don't advise the amateur to do his own sharpening—take it to a professional who makes a business of it.

Amateurs are noted for their dull saws, and they seldom know *when* they are dull. A saw gets dull sooner than you might think, and nothing causes poor joinery and ill-fitting joints more than a dull saw. When a saw gets dull it also loses some of its set, and this causes the saw to bind and to run. By "run" we mean cutting to a gradual curve, one side or the other, away from a straight line, and a saw that is improperly set will run in spite of anything you can do.

After a saw is filed, it is set, with a setting tool, and it takes considerable experience to give it the right degree of set. As a final step, a professional will stroke the sides of the teeth with a hone, lightly, to remove the slight burr that causes rough cuts. So to sum it all up, have your saws filed and set frequently, by a professional, and take care of them. Be careful where you lay a saw, and never let the teeth touch anything harder than wood. Keep an oil-can handy, filled with a 50-50 mixture of oil and kerosene. Give each side of the saw a squirt when starting a long cut or in hard wood, and it will prevent binding.

Planes

Planes for boat work are about the same as those used for house carpentry, except for a couple of types used by professional boatbuilders—the "bullnose" plane for working close into corners, and hollow-and-round planes for smoothing up planking. For the amateur, at least two planes are a must, a 9 in. smooth plane for general use, and a 6 in. block plane, for cutting across the grain and trimming off corners. To be completely equipped, however, you should have a fore plane, about 18 in., or a jointer, 21 in. These long-soled planes are needed for dressing planking or any long lumber where a true, level edge is required. The length of the sole reduces the possibility of hollows and high spots. A rabbet plane is practically a must, since rabbeting is required frequently in deck and cabin joinery.

Old time craftsmen would never use anything but wooden-sole planes, or planes whose whole body was wood. The reason is that they would cut smoothly on any kind of wood without sticking or chattering, and with little or no friction. Iron sole planes drag and stick badly if the wood is green, or full of pitch, as in yellow pine, and will chatter and jump on very hard wood or cross grain. Today, the old time wooden plane cannot be had, except at country auctions, or if you happen to know an old timer who is retiring. I think it's a danged shame, for once you've used one, an iron plane will seem like a very poor tool. Such is "progress."

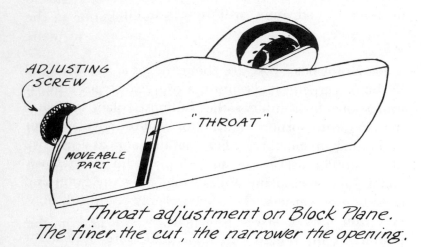

*Throat adjustment on Block Plane.
The finer the cut, the narrower the opening.*

Iron planes have smooth or corrugated bottoms, and the corrugated type is supposed to have less friction, which is logical. All planes should have caps, or double irons.

There are right and wrong ways to use a plane. The right hand does the pushing, and keeps the plane level athwartships. The left hand serves as a guide to hold her on course. In planning the *edge* of a plank, the tips of the fingers of the left hand extend below the plane to keep the plane centered. Strokes should be as long as is convenient without stretching your arms to the limit—don't jab with the plane. When the last stroke at the end of the plank is made, don't let the plane tip over the end, but shoot it out in a straight line. On the return stroke, don't drag it back—rather, lift the rear of the plane off the work. In all planing except cross-grain, the plane should be held straight, with the cutting edge at right angles to the direction of stroke. Always note the direction of the grain before starting to plane—it is impossible to plane smoothly against the grain.

For accurate fitting it is generally necessary to plane off the end grain. For this a block plane is used. Block planes must be very sharp, and the bit set for a very fine cut. Hold the plane bottom up, against the light and sight along the sole. The cutting edge should be just barely visible. Hold the plane at an angle of about 30 degrees to the direction of stroke. Lift the front of the plane before the cutting edge runs off the far edge, otherwise you'll splinter it off. Then reverse and cut from the opposite side in the same manner.

Sharpening and setting plane bits requires skill and accuracy. First a word about the oilstone, which should be a natural stone—not an artificial abrasive. DON'T use ordinary lubricating oil on it, or grease or water. Use only very light machine oil, or a 50-50 mix of lubricating oil and kerosene—otherwise the stone will become glazed very rapidly. Wipe it clean when through and keep it covered when not in use.

Put a few drops of oil on the stone, and be sure the stone rests solidly and level. Grasp the bit with both hands, with the fingers, as many as there is room for, close to the cutting edge. Place the blade on the stone slightly askew, and at the proper bevel, and rub it back and forth. The proper bevel for plane bits is from 20 to 25 degrees. Don't rock the bit—bear down firmly and steadily, and do your level best to maintain the exact bevel. I was always taught to move the bit in a slightly rotary fashion, rather than straight forward and back. This eventually gives a slight curvature to the cutting edge, which prevents either of the corners of the bit from scoring or marking the wood.

The final act in sharpening is to remove the burr. Place the bit face down, bevel side up, and slide it back and forth on the stone a couple of times. Now if your bit is really sharp, it should nick your fingernail at the slightest touch. One old timer I used to know would try it on his cheek—he said if it would cut whiskers it was sharp enough for wood.

The amount of set, or the amount of cutting edge exposed below the cap, is very important. I would recommend about 1/16 in. for coarse cutting, or on very soft woods, and 1/32 in. for fine planing, and on hard woods. When your plane jams up with chewed-up shavings it is a sign that either your set is wrong or you are planing cross grain, improperly.

A squirt of that kerosene-and-oil mixture on the bottom of the iron plane from time to time will help reduce drag and sticking. Always lay your plane on its side when you put it down, to protect the cutting edge.

Chisels

For amateurs, I would say the minimum number required are three, a ½ in., a 1 in., and a 2 in., but I strongly recommend acquiring a full set, which would be six or eight, starting with a ¼ in. It is surprising how often chisels are needed, and when a ⅜ in. chisel is needed, a ½ in. won't do.

There are two types of chisels—socket and tang. Stay away from the tang type, in which the tang of the blade is driven into the handle. They are characteristic of cheap, bargain-counter tools, and are not suitable for either amateur or professional. Buy only socket butt chisels, and if the handle splits or mushrooms from using a mallet, a new one can easily be fitted.

Chisels are sharpened just like plane bits, and the bevel is the same—20 to 25 degrees. The smaller the angle of bevel, the better it is for thin, precision cutting. The greater angle is better for heavy, deep cutting in hard woods. Here's a tip on using a chisel correctly: when using a mallet, *don't* hold the chisel by the handle, hold the blade near the cutting edge. Thus your hand can rest on the work and guide the cutting accurately. This is especially important when cutting mortises.

I mention later, in my remarks on bunging over fastenings, the use of a "slick," which is a heavy chisel not less than 2 in. wide. This chisel is a must. Actually, with experience, you can do many things with it. My advice is to buy the biggest and heaviest one you can. Because of its weight, size and balance, you can use it like a plane or a drawknife, and you don't need a mallet. When planking, it's an excellent tool for taking down the plank-edge quickly to rough shape. Just hold it flat

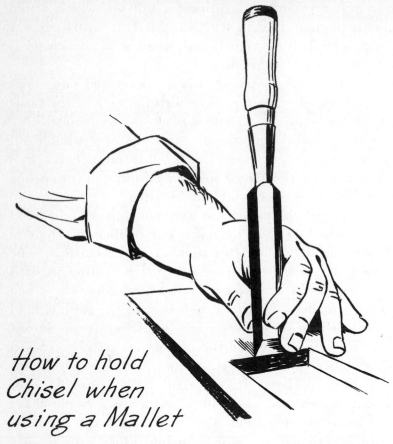

How to hold Chisel when using a Mallet

to the work and shove it along. An experienced man can take a shaving off soft wood, such as pine or cedar, just as neatly as with a plane, and in places where a plane can't be used. In framing or timber work, the slick is very useful for shaping and truing up a member *after* it is in place and fastened off.

In addition to the chisels I have mentioned, you should have a couple of gouges, say a ½ in. and a 1 in. They are needed, among other things, for cutting clearance or mortises for drift and bolt heads.

As a final word on using chisels, don't take too big a bite—small chips are the mark of a professional. In cutting a mortise, for example, most amateurs wallop the handle with the mallet as though they were driving a spike, driving the chisel so deep it can't be freed, let alone remove the chips. The reason for cutting *lightly* is that you have better control; you won't cut where you shouldn't, and you won't split the wood.

Boring tools

Naturally, you need a brace and bits. The brace should have a reversible ratchet, ball-bearing chuck and head, and a 10 in. sweep. Auger bits can be had in sets of from eight to fifteen, but I certainly would recommend nothing less than a complete set, from ¼ in. to 1¼ in., by sixteenths. Sooner or later you'll have need for every one of them.

While not indispensable, an expansive bit is a handy thing to have, particularly when installing hardware. They come in several sizes, for boring shallow holes from ⅝ in. to 3 in. in diameter.

Although they are not boring tools, the following items are used in bit braces, as well as in hand and electric drills. First of all, you should have a rose bit, which is a countersinking tool. In addition you must have the various patented countersink-counterbore tools, made for hand or electric drills, which I discuss at some length under FASTENINGS. These had best be obtained from a marine supply house, where you can choose the type that suits your fancy, for the specific size and type of fastenings you are using.

For boring for drifts you must have ship augers, barefoot, or without worms. For the small boat normally

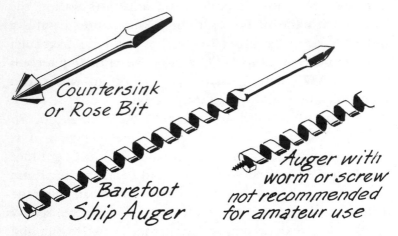

Countersink or Rose Bit

Barefoot Ship Auger

Auger with worm or screw not recommended for amateur use

associated with amateur construction, ⅜ in. and ½ in. are the sizes most likely to be needed.

A good, professional-type hand drill is also a must. Get one with double-pinion gears, if you can, for easier operation. For drills, you have your choice of carbon or high-speed steel, the latter being the more expensive, but required for high speed drilling in metals. I urge you to have a full set of drills, either "wire gauge" or twist, so that you can bore accurately for every size and type of fastening. To go with the drills, get a drill index, which is a rectangular piece of metal full of holes. Drop a screw into the hole that fits, and you automatically read the proper drill size for the shank. A good set of drills with a drill index might cost in the neighborhood of $10 or $12, but it would be money well spent, and since you will be working mainly with wood, they'll last a long, long time.

Screwdrivers

Hand screwdrivers should be high-quality, carefully tempered tool steel, of a nationally-advertised make, such as Stanley or Millers Falls. I would advise having a complete set—¼, 5/16, ⅜, and ½ in. to fit every size screw. The blade of the screwdriver must fit the slot of the screw exactly, both in width and in thickness. Unless the tip of the blade is square, straight and sharped-edge, it is liable to slip and mar the wood.

Where many screws are to be driven, the automatic, or "push-pull" screwdrivers are great time-savers. Get the largest and heaviest you can buy, and in using it, always be careful to see that it is seated accurately. One slip, and you have a bad gouge in the plank.

For real heavy duty work, when you are driving #14 screws or larger, screwdriver bits and a brace are recommended. These bits come in sets of four, from ¼ to ½ inches wide. The brace gives you great leverage, and the screw can be driven home smoothly, in one continuous action.

Always lubricate the threads of a screw before driving, particularly when working with oak. The best lubricant is the old-fashioned yellow laundry soap. Just drag the screw across the cake and the threads will pick up the right amount. One point to remember is that once you start to drive a screw, *don't stop until it is driven home.* A heavy screw being driven into hard wood will "freeze" the instant you stop turning.

Measuring, scribing and spiling tools

These are the rules, squares, bevels, and compasses. Of rules, it's a matter of personal preference whether you

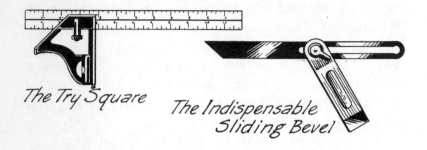

The Try Square   *The Indispensable Sliding Bevel*

have the folding wood type or the push-pull, retracting steel type. Each has its good points, but the steel ones are valuable in boat work for measuring "inside-to-outside." Whichever you use, it should be a 6-foot one. House carpenters' 2-foot folding wood rules are of no use in boat work. For lofting, framing and layout work when building, a steel tape measure is a must—cloth tapes are useless.

Two squares are necessary—a 24 in. by 16 in. common steel square, and a 12 in. combination square, which can be used as an inside or outside try-square, a depth gauge, or a rule.

A good bevel is the most important tool of the boat

builder, outside of the hammer and saw. *It will be used on practically every piece of wood that goes into the boat,* for it is an old saying that "nothing is square and everything is beveled." So get a good sliding "T" bevel, with a blade not less than 8 in. long.

For scribing, spiling and making templates you must have a 6 in. carpenter's pencil compass, and a pair of 8 in. wing dividers.

Clamps

Let's face it—you can't build or repair boats without clamps of some sort, and there are but few amateurs who ever have enough of them. The common C-clamps, in sizes large enough to be useful, are expensive. An 8 in. clamp, for instance, will cost at least $2.00, and the larger sizes will run as high as $10.00. So you can see why the non-professional gets along with as few as possible. If you are building a boat you really should have at least half a dozen, and the bigger the boat the more you need. Gluing up a hollow box spar requires from 30 upwards.

Like other amateurs with whom boat work is a constant, continuing interest, I have accumulated through the years a dozen or more clamps of various sizes. But when I need many more I canvass all my boat-minded friends, and I have no difficulty in rounding up all I need. That's one of the nicest things about boat-minded amateurs—they are so conscious of their own difficulties and painful experiences that they will always bend a sympathetic ear, and freely give of what they have.

Along with as many C-clamps as you can acquire, a bar-clamp is very handy. This can be either the steel-

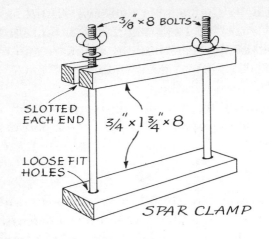

3/8" × 8 BOLTS

SLOTTED EACH END

3/4" × 1 3/4" × 8

LOOSE FIT HOLES

SPAR CLAMP

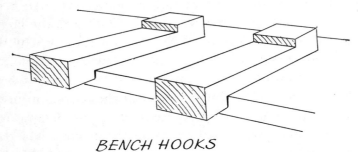

BENCH HOOKS

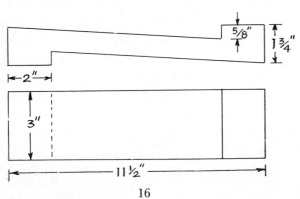

5/8"

1 3/4"

2"

3"

11 1/2"

16

bar or the pipe style. They are sometimes called "gluing clamps," and are useful principally in interior cabinetry, paneling, making doors, and other units. However, you *can* get along without one, so it is not a must.

The above comprise all those hand tools I felt warranted specific comment. There are many other minor tools used occasionally, as your own common sense will tell you. Naturally, you'll need such things as nailsets, a hacksaw, files, rasps, punches, and if you know how to handle them, a good draw-knife and a spoke shave. These last are very handy for shaping planks, after spiling, and for finishing inside-curves and rounding off sharp edges of interior joinery.

In the final analysis, your collection of tools is as extensive as your specific needs require, and your circumstances permit. A 14-foot outboard "kit" boat can be put together in a workmanlike manner with a very few common tools, and the same is true of minor repairs or alterations. But if you're starting from scratch to build a 36-foot sport fisherman or a 25-foot sloop, your kit of tools must naturally be nearly as complete as that of the professional boat builder. As I said before, you rely on your common sense, and what you lack you either buy or borrow.

Here are two tools you can make yourself. The first is a wooden clamp designed for building hollow box spars, and you can make as many as you need at a nominal cost. As the illustration shows, it consists of two wood cross-pieces and a pair of common ⅜" carriage or machine bolts with wing nuts. Dimensions are only typical, and should be determined by the size of the spar you are making. The wooden pieces can be made of scrap lumber, so long as it is reasonably hard. Notice that the upper piece is slotted at the ends to take the bolts. This is done so that you can put the clamp together very quickly. Remember, you'll be working with waterproof glue that has a very short working time, and speed is of the essence. A 30-foot spar might require some 40 clamps, so this means setting up 80 wing nuts, not to mention assembling. You can't do it single-handed, and you will need several willing helpers who are more or less ambidextrous!

Next is a *bench-hook,* always found in the shop of any old-time craftsman. A pair of them hooked over the front edge of the work bench are mighty handy when working with a chisel on small parts, to hold the work without clamps when mortising, cutting rabbets, etc. They should be made of hard, dense wood, such as maple, locust or white oak. When finished, give them a good soaking in hot linseed oil, and let them dry for a week or two. Once you have used them I'll warrant you'll take a good deal of pride in them, for they serve a purpose nothing else can fulfill.

POWER TOOLS

Aside from their importance in labor-saving, power tools are a tremendous aid for the amateur in that their mechanical precision makes up for his lack of skill with hand tools. Few amateurs can rip a plank straight down a line with a hand saw as accurately as a bench saw, nor dress a plank edge with a fore plane as true as a power jointer. Indeed, power tools are directly responsible for the present high degree of craftsmanship that has been attained by so many amateurs. The following list covers those most needed for boat work.

The Bandsaw

A good 12-inch bandsaw taking stock up to 4 inches is adequate for amateur work, and can be had for under $100. You should have several spare blades of different sizes. A ¼-inch blade can cut to a small radius, but when cutting a plank to a slow curve a wider blade gives a truer cut, and has less tendency to run as the density of the wood varies. 3-wheel saws are not desirable, as the blade is flexed more severely and soon loses its temper.

The Bench Saw

This should be the tilting arbor type, with 8 or 9-inch blade. You'll need rip, crosscut and combination blades, carbide tipped. There is also a blade especially designed for cutting plywood. A set of dado blades adds to the versatility of your saw for rabbeting and spar making.

A word of caution—all power tools are dangerous if used improperly. *Never* set the blade higher than ¹⁄₁₆-inch plus the thickness of the stock you are cutting. This reduces possibility of a kick-back.

The Electric Handsaw

This is a most useful tool by virtue of its portability. The 7-inch, 1½ h.p. size is recommended. Be sure it has a cutout on the blade housing so you can see the cutting line, and an anti-kickback safety clutch.

The Sabre Saw

This serves as a portable bandsaw, and can be used in corners or small quarters where there isn't room for the handsaw. In working with plywood it is almost indispensable. It should have a 1-inch cutting stroke and circle-cutting and edge-guide attachment. A set of 7 or 8 different blades are available for cutting various materials, including metal.

The Jointer-Planer

While not indispensable, a 4-inch jointer is a valuable tool for precision joinery. You can dress a plank edge absolutely true, sharp and square. The depth of cut can be set to take off a very fine shaving, which is mighty important for accurate fitting. It will also cut rabbets up to $\frac{3}{8}$-inch deep.

The Electric Drill

A good medium or heavy-duty drill is a necessity, since practically all fastenings must be bored for. The cheap, low-powered $\frac{1}{4}$-inch jobs so widely exploited are a waste of money. With the innumerable deep-boring jobs in dense woods frequently encountered, you need a $\frac{3}{8}$ or $\frac{1}{2}$-inch drill, with a $\frac{3}{8}$ h.p. motor. The variable-speed type has greater versatility.

You'll need a good set of high-speed drills and a drill index. A most useful accessory is a screwdriver and speed-reduction kit.

Electric Sanders

Three types are available—the belt sander, the orbital sander, and the disk sander. Each is best for a particular job, and all have their limitations.

The belt sander is a heavy, powerful, fast-cutting tool suitable for large, flat areas such as decks and cabin roofs. It stands in a straight line, and cuts right up to a vertical member. For sanding vertically and on hull planking, it is too heavy and tiring to hold, and no good on curved surfaces.

The Orbital sanders are light in weight, for finished, rather than rough sanding. Since they cut in a circular motion, you can't sand close to a vertical member, and if the part being sanded is to be finished bright, unsightly swirls will appear when it is varnished. A more useful version is the "dual" sander—a flick of a switch and it changes from orbital to straight-line sanding, which is necessary for sanding with the grain. Another advantage of this model is that you can sand right into a square corner. All things considered, it is the most practical general-purpose sander.

Disk sanders are most useful for fairing up uneven planking, or rough sanding irregular surfaces. They cut fast, but as with the orbital type, leave circular swirls. As an improved version, there is the "flex-pad," a flexible rubber disk for use in an electric drill. Sanding disks are affixed to the pad with an adhesive. This can be used on any member, whether convex or concave.

II

Woods for Boat Work

It is seldom that the amateur has more than a general knowledge of woods—their characteristics, availability, and their suitability for boat work; nor is there any reason to expect him to have the professional's understanding of the subject. But you can't use just any old piece of lumber in boat work. Not all woods are suitable. They must be selected for certain specific characteristics —their workability, strength, resistance to rot, and ability to take a finish.

When the average non-professional has need for wood for boat work, he will all too often make a bee-line for the nearest building supply dealer or lumber yard. What he comes out with is some kiln-dried lumber, whose life has been cooked out in the drying process, whose natural oils have been leached away, and which is naturally receptive to dry rot. Kiln-dried lumber is *dead,* as compared with air-dried, and it *not* suitable for boat use—if you want a first class job.

So where, you may ask, do you obtain boat lumber, air-dried and selected expressly for boats? Along the seaboard, or wherever there are boat building yards, you will find lumber suppliers nearby. Some advertise in the yachting publications. If you need but a small amount, many boatyards will supply you from their stock; in fact many look upon this business as a desirable source of added income. While you may have to pay a slight premium at this source, it is generally worth it for the convenience and the assurance that you are getting suitable wood.

The seasoning of boat lumber is accomplished by racking up the green stock, sometimes under cover, and allowing it to dry out naturally in the open air. This may take a year or more, depending on the species and the size of the timber. The timber ends are generally painted to prevent checking.

There is one exception to the above which should be noted. Oak for steam-bent frames should be *green* stock, rather than seasoned, since it can be bent more easily, and is less inclined to become brittle in the steaming process.

The list which follows includes all those native woods which are recognized as suitable for boat work. I have endeavored to include all of those properties which I deem pertinent to their use—strength, hardness, durability, workability, resistance to rot, and usefulness.

Ash

Commercial white ash rates high in strength, hardness, durability and resistance to shock. Its workability and resistance to rot is about the same as oak. Its use is confined to small boats or light construction, mainly for steam-bent frames where the least weight with maximum strength is desired. Oars and boat-hook handles are commonly made of ash.

Cedar—Alaska, Port Orford, and Western Red

I have grouped these West coast cedars together because they have similar characteristics. They are light in weight, rather soft and moderately low in strength. Available in clear, fine, straight grain and free of knots, they are excellent for planking and cabin joinery. They are easy to work with, take paint or varnished finish well, and are highly resistant to rot. Being a soft wood, cedar is easily marred or punctured by sudden shock. Fastenings at plank ends should be made with care, or splitting will occur. Of the three, Western Red Cedar is the softest and the weakest, and therefore somewhat less desirable.

Alaska yellow cedar is being used more and more as its superior qualities become known. It it the strongest, hardest, and heaviest of all the cedars, with the same high resistance to rot. The grain is uniformly straight and clear, and it is easily recognized by its bright yellowish color. It can be had in long lengths and free of knots. For planking it is the best of the cedars.

Port Orford cedar is a light brown in color, rather coarse grained, and even textured. It is easy to work, but splits easily when fastening plank ends. It comes in long lengths and is light in weight, hence is very popular for planking where weight is important.

Cedar—White

White cedar includes two species, northern, which grows in the Lake States and the northeastern sections, and southern, which grows near the coast in the eastern States. About the only difference between the two is that the northern white cedar is a relatively small tree. The wood is lighter in weight, softer, and slightly weaker than the Western cedars. It has a fine, uniform texture and a very aromatic odor, is easily worked and is highly resistant to rot. White cedar is used almost exclusively for planking, can be steam-bent easily, and is the lightest planking material known.

Unlike the Western cedars, white cedar is rarely obtainable free of knots. This is not a serious fault, however. Small, tight knots are not objectionable, and do not appreciably affect its strength. Large loose knots are driven out, and the holes are plugged with the same cedar, and this is standard practice everywhere. Cedar swells easily and quickly when wet, a desirable quality in planking.

Cherry

Native wild, or black cherry is an excellent wood for boat work, but the supply is rather limited, and therefore is sometimes difficult to obtain. The wood is strong, quite hard, shock-resistant and fairly easy to work. It is quite resistant to rot. The grain is fine, uniform and close. Color varies from light red to dark reddish brown, and varnish brings out a beautiful luster that is similar to fine mahogany or madeira. It takes a truly handsome finish. Cherry is what I call a stable wood, in that it stays in place if well seasoned, without shrinking, warping or checking.

Cherry's rather gnarled, rugged habit of growth sup-

plies many natural-grown crooks or knees. Therefore it
is an excellent source for sawn frames, knees, breast-
hooks or any structural members that require considerable curvature in the natural grain. Its beautiful finish
when varnished makes it very desirable for interior trim,
joinery and furniture. It does not respond to steam bending too well.

Cypress

Southern cypress has one characteristic for which it is
universally acclaimed—very high resistance to rot. Indeed, it is often called "the wood eternal." It contains
a natural oil or resin that acts as a preservative. The
wood is fairly light, moderately strong, and moderately
soft. It holds paint well, but is not suitable for a varnished finish. When painted, the grain is difficult to
hide, and will invariably show conspicuously through
numerous coats. It is available in wide boards, long
lengths and clear, straight grain. It might be termed a
general utility wood, since it is used for planking, decking, exterior and inside joinery.

Cypress has one undesirable peculiarity. It soaks up
water to a surprising degree. While it has been used for
planking boats for countless years, particularly in the
Southern States, it is not considered an ideal material for
the purpose—such boats are inordinately heavy due to the
very high absorption. This is particularly noticeable in
small boats, where weight is an important factor. A row-
boat planked with cypress will actually seem waterlogged,
when compared to one planked with cedar.

Elm

Rock elm is heavy, hard, and strong, with high resistance to shock, and to rot. It is noted for its ability to withstand extreme deformation without failure, and it can be steamed to very sharp bends. Because of this attribute, and its great strength, it is used primarily for frames in small craft of light construction, such as canoes and rowing skiffs. It is also used infrequently for lap-strake planking. It is more commonly used in England than in this country, particularly in light sailing craft.

Elm is a stable wood—it stays put without warping or checking. Its grain is close, fine, and straight. Unfortunately, it is not obtainable everywhere, and it takes a bit of searching to locate the nearest source of supply.

Rock elm is grown mainly in the North Central States, and should not be confused with the common American elm or the slippery elm, neither of which are suitable for boatwork, since they are softer and much lower in strength.

Fir

Douglas fir is quite strong, moderately hard, and heavy. Its grain is somewhat coarse, open pored and generally uniform. It has a tendency to check badly, and split in the way of fastenings. It absorbs water readily and is low in resistance to dry rot or decay. While it is highly regarded for structural timbers, its practical value in boat work is limited generally to bilge stringers, clamps, shelves and spars. Strip decks are often made of fir. Vast quantities of fir are employed in making plywood, which will be discussed under a separate heading.

Hackmatack

This wood is also referred to as larch or tamarack. Its principle value is for natural-grown knees, where it is generally preferred over other woods. It is heavy, strong, close-grained and resistant to rot. While its color may often vary from light to dark brown, it takes a varnish finish nicely. It is generally obtainable only from suppliers who specialize in woods for boat builders.

Locust

This is one of our hardest and heaviest woods, and highly regarded for boat work. It is exceedingly strong, close-grained and tough. It seldom checks, warps, or shrinks, and is very difficult to split. Dry rot rarely attacks it. It is the ideal wood for bitts, cleats, tree-nails, wedges, tillers, strongbacks, or any structural member where great strength and rigidity are required. Because of its density and close grain it is rather difficult to work, which means that sharp, heavy tools are needed. This is one wood that responds to the use of a wood rasp for rough-shaping, without splintering. Its natural golden yellow color is handsome when varnished.

Locust should be selected with care, for the trees are often attacked by borers, which leave sizable holes or tunnels.

Oak, White and Red

For hundreds of years this noble wood was been a symbol of strength and longevity the world over, and no substitute has yet lessened its popularity in ship and boat building. Its superior qualities of toughness and durability are too well known to require elaboration here.

White oak is preferable to red oak for several reasons. Its cells or pores are plugged with a growth called *tyloses,* which means that the wood can absorb only a limited amount of water, and swelling is reduced to a minimum. Red oak does not have such a structure, and soaks up water to a much greater degree. This makes it softer, weaker, and easily damaged by shock and abrasion. Then too, white oak is high in resistance to rot—much more so than red oak.

As to sheer strength and ability to withstand severe distortion, white oak rates somewhat higher than red oak. For steam bending white oak is considered superior—and it should be *green* wood, *not* air dried and well seasoned.

Oak holds paint well, but if a varnished finish is contemplated there is one fact to remember. Oak has the unfortunate habit of turning black when wet, and it will happen wherever the varnish is worn or weathered off. This means keeping at it all the time with the varnish. Below decks, of course, you don't have this problem to contend with, but outside where the weather does its dirty work it is quite a chore to keep a bright finish on oak. Once the black stain appears there is nothing you can do but remove all the finish and bleach the wood with oxalic acid, a mean job that is not always effective.

To sum up, for all framing and structural members, use white oak if you can, red oak if you must. Fortunately it is in plentiful supply and obtainable almost everywhere.

Pine, White

There are a number of varieties of white pine on the market and all have been used in boat work, but the best are the Northern white pine, common to the Eastern States, Western, (or Idaho,) and the Sugar pine of the West Coast. Ponderosa pine is slightly less desirable, and Norway pine is very inferior—actually, it belongs to the yellow pine family.

White pine is known for its creamy color, soft, uniform texture, and the ease with which it can be worked. It is usually straight-grained, light in weight, moderately low in strength, and holds paint well. Where conditions are favorable to dry rot, it is moderately low in resistance.

White pine is excellent for interior trim and joinery. In some localities it is used for planking, I suspect mainly because of expediency or low cost. In fact, it is sometimes referred to as "Maine mahogany." While stronger than cedar, it absorbs water easily, and a boat planked with white pine will be a heavy boat. It does not respond to steam bending very well, nor does it swell or shrink as much as some other woods. It is never finished bright except below deck, where it is sometimes given a natural finish of clear stain and wax.

Pine, Southern Yellow

Long leaf yellow pine is admirably suited for boat work, since it is moderately low in cost, and is obtainable in very long lengths with a clear, straight grain free of knots. The wood is dense, quite strong, hard and durable. Quarter-sawn, or edge grain should be specified.

and it should be carefully selected. It varies considerably in denseness and "fattiness," or resin content. Pitch pockets are common and should be avoided as far as possible. Pick the wood that seems to have the least amount of pitch, since it exudes under a hot sun and will bleed through paint. In fact, one of the faults of yellow pine is that it does not hold paint very well.

Due to the resin or pitch it contains, yellow pine is heavy, (slightly less than oak), and does not readily absorb water. In warm, or tropical waters, it is superior to oak for keels, since it is less susceptible to marine borers and rot.

Yellow pine is used for planking in commercial craft, but in pleasure boats generally it is used for the sheerstrake only. Its other uses are floor-timbers, sawn frames, engine beds, bilge stringers, clamps, shelves, log rails or bulwarks, centerboard cases and centerboards.

Yellow pine is sometimes used for strip decks, but it should be quarter-sawn edge grain, and not too "fatty." Short-leaf yellow pine, commonly called N.C. pine is not considered as desirable for boat work. It is lower in density, lighter in weight, and not as strong.

Pine, Oregon

This is not a pine at all, but simply another name for Douglas fir. The only reason for mentioning it is that you will often see it specified—mainly for strip decks. For this purpose it is preferable to yellow pine.

Redwood

In recent years the use of redwood in boats has increased, although it is of relatively minor importance in the list of boat building woods. Redwood is moderately light in weight and in strength, and it is generally very straight and evenly grained. Its principal value for boats is in interior trim and cabin finish. It holds paint well and takes a handsome, though rather dark finish. Its greatest virtue is its very high resistance to rot. Its main fault is its low resistance to shock. It is brittle, and does not respond to steam bending.

Spruce, Sitka

This is the favored wood for spar making. Light in weight, moderately hard, stiff and tough, it is obtainable in very long lengths with straight grain, uniform texture, and free of knots. It is very high in strength for its weight, warps and shrinks very little, and is easy to work. It is not too resistant to rot, and therefore should be sealed to prevent absorption of water.

Spruce, Eastern, or White

Eastern spruce has been used by shipbuilders in New England and the Maritime Provinces of Canada for several hundred years. It has been used for planking, decking, and keels in commercial craft. For pleasure boats it is used in decking, rarely for planking, but never for keels. Its weight, strength and resistance to rot are about the same as Sitka spruce. It is finer, closer grained, and a silvery white in color, while Sitka is coarser and yellow.

The principle objection to Eastern spruce is that it generally has many knots, and is not obtainable in long lengths. However, it is my personal opinion that it is an excellent substitute for Sitka at a much lower cost, *providing* you carefully select it for clear stock with a minimum number of small, tight knots. This generally means overhauling several hundred planks to get one that is clear, but it can be done.

Walnut

Black walnut is an excellent wood for interior trim, much prized for its figured grain and fine finishing qualities. Because of its dark color it should be used with restraint, as an accent to painted areas. Large areas of it tend to make a cabin dark and gloomy. The wood is heavy, dense and strong, with a minimum of shrinking or swelling. It does not lend itself to steam bending.

WEIGHTS OF WOODS

The following table gives the weight per *cubic foot* of each of the species previously described, as supplied by the Forest Products Laboratory. The weights are based on air-dry specimens having a moisture content of 12%. Remember, they are only *average* weights, and

Species	Lb.	Species	Lb.
Ash	41	Hackmatack	36
Cedar, Alaska	31	Locust	48
Port Orford	29	Oak, white	47
Western	23	red	44
Northern white	22	Pine, white	25
Southern white	23	yellow	41
Cherry	35	Redwood	28
Cypress	32	Spruce, Sitka	28
Elm	35	Eastern white	28
Fir, Douglas	27	Walnut	38

individual samples may vary as much as 10% from the
figures shown. However, their main usefulness is for
comparison of the various species.

IMPORTED WOODS

There are but few species of foreign woods imported
in any considerable amount for boat building purposes,
and they are the various Mahoganies and Teak. Small
quantities of Madeira or Horse-flesh are brought in from
the Bahamas and the West Indies for natural-grown
knees and frames for small craft, but their cost is high
and they are difficult to obtain. Rare and exotic woods,
prized for their decorative value, and correspondingly
expensive, find their way into some of our finest yachts.
But the average amateur had best think in terms of practical usage, and stick to woods that are easy to obtain at
a cost comparable to our domestic woods.

Mahogany, once used only for its decorative appearance as trim, is now used for practically all parts of a boat
except keels and frames. Once difficult to obtain and
expensive, it can now be had wherever boat lumber is
sold, and at a reasonable cost. It is high in strength,
tough, durable and resists rot. It has a minimum of
shrinkage, warp and swelling, is exceedingly close grained
and is difficult to split. Paint sticks to it agreeably, and
its beauty when stained, filled and varnished is unsur-
passed. It is relatively easy to work with and has the
density to hold fastenings securely.

Because of its resistance to shock, and its high strength
in relation to its moderate weight, it is unequalled for
planking. Rails and rail caps and coamings, and all ex-

posed trim that really takes a beating in normal use, will stand the wear and tear with a minimum of depreciation if made of mahogany. Finally, varnished mahogany traditionally *belongs* in pleasure craft, for it enhances both their beauty and their value.

Not all mahogany is alike, however, for there are several kinds grown in different countries—each having its own characteristic color, grain, density and strength. To use these fine woods intelligently you must know wherein they are different and what you are buying. Most amateurs simply ask for "mahogany" and take what is offered, not realizing that there are several kinds to choose from. In making alterations in a boat, you can't match up the color and grain of the existing mahogany members unless you can recognize the species and order accordingly.

There are but three varieties of *genuine mahogany* on the market in commercial quantities and readily obtainable. They are the African, Honduras, and Peruvian.

African Mahogany

Botanically *Khaya,* this is obtained principally from the Gold Coast, the Ivory Coast, and Nigeria. The trees are noted for their immense size, and yield longer and wider lumber than any other type. The logs measure from three to six feet in thickness, and fourteen to thirty-six feet in length. For this reason, as well as its lavishly varied figure and grain, a large amount of African mahogany goes into veneers and plywood.

When freshly cut, the wood is a salmon pink in color, which changes when exposed to light and air to a pale, reddish brown. One distinguishing feature is seen when you look at the end grain—the concentric growth lines, or annual rings are totally lacking, which is true of no other mahogany. Of the three varieties, African is the softest, lightest, least strong, and has a milder texture and slightly larger pores. Its color generally runs more to the reddish tones than the others. Its weight will average about 32 lbs. per cubic foot.

As to its uses in boat work, its principal merit is in the infinite variety of its figured grain, ranging from the simple, straight stripe to the rich and complex figures of mottle, crotch or burl, and swirl. Hence it is used for interior cabinetry where its decorative beauty can be displayed. Without a doubt, it is the most beautiful of all the varieties.

Honduras Mahogany

Botanically *Swietenia,* this is the trade name or classification for all the Tropical American mahoganies grown on the mainland of Central and South America from Southern Mexico to Colombia and Venezuela, including all the Central American countries.

It is heavier, slightly harder and stronger than African, and is lighter in color, running more to a light orange than reddish. Of the three varieties, Honduras is more commonly used in boat work, and is the easiest to obtain, and the lowest in cost. It weighs about 36 lbs. per cubic foot. It is generally straight-grained and takes a handsome finish.

Peruvian Mahogany

Of the same genus, *Swietenia,* Peruvian mahogany is known as "Aguano" in South America. It is a relatively new find, having been discovered about 25 years ago in a vast area around the tributaries of the upper Amazon, in a part of Ecuador, Peru, Bolivia and Brazil. It is far from the shipping points on the Pacific coast, and is gotten out of the jungle with considerable difficulty. Hence it is the most expensive of the true mahoganies.

It is also the *best*. It is the heaviest, about 38 lbs. per cubic foot, and is harder, denser, and darker in color than Honduras. It is quite brown in tone, with little of the orange or red of the other varieties. It takes a beautiful finish, and is straight grained, with great depth of color.

Peruvian mahogany is really tough and hard—in fact it is almost twice as hard as African. Its fibers interlock in such a manner that it is extremely difficult to split. All in all, its relative superiority where fine wood is called for easily justifies its somewhat higher cost.

Philippine "Mahogany"

Let's start by clearly stating that this is *not mahogany*, in fact it is not even related to the true mahoganies. It is a popular trade name for at least three different species of trees native to the Philippine Islands. It bears some resemblance to mahogany, and that probably accounts for the origin of its misnomer.

There is probably more Philippine mahogany used today for boat planking than any other wood, which in itself is a pretty good recommendation. It is moderately hard, moderately strong, normally straight grained, and highly resistant to rot. It is lighter in weight than the true mahoganies. Its working qualities are excellent, it warps and swells very little, and seldom splits at end-fastenings. It responds to steam bending, and has a moderate resistance to shock.

As compared to true mahogany, this wood finishes rather poorly. It is open-grained and rather coarse, and doesn't have the same depth of color or eye-appeal. Manufacturers of stock boats use it not only for planking, but for trim, cabin joinery and bright-work, because of its comparatively low cost. But even with the judicious use of fillers and stains, it rarely attains the rich appearance of true mahogany.

It is marketed in two groups—*light* red and *dark* red, and for boat work you should specify "quarter-sawn straight grain, First and Seconds." Light red is preferable for planking, and dark red for finishing bright.

Teak

Of all the woods used in boat and shipbuilding, teak has no peer for durability, for it has inherent qualities not found in any other wood. Exposed nakedly to the elements, with no protecting film of paint or varnish, it will endure for countless years without change.

Native to India, Burma, and various islands in the East Indies, teak is considerably more expensive than any other wood used in boat work. In fact, its cost is the only factor limiting its use, and one generally associates it with work of the highest quality. This does not mean that it is beyond the means of the average boat owner. Many small boats are greatly enhanced in appearance, longevity and value by a relatively small amount of teak used judiciously where it counts the most. A 24-foot

skiff can have her decks or cockpit floor of teak without
raising the total cost disproportionately. In the first
instance, an open skiff has very little deck area, so little
material is needed. The cockpit floor can be double-
planked—⅜ in. teak laid over fir playwood, and again, a
relatively small amount is required. So if you have a
champagne taste and a port wine pocketbook, don't let
the price of teak scare you.

Teak is very hard, dense and heavy. It weighs about
47 lbs. per cubic foot. Most of its fine qualities are due
to a characteristic possessed by no other wood, as far as I
know . . . its pores are filled with a peculiar resinous oil
so potent that it will not rot, cannot absorb water, and is
such a natural preservative that it needs neither paint nor
varnish. Teak is relatively difficult to work, and it will
dull sharp tools quickly. Whether it is due to the resin
it contains, I do not know, but it will take the edge off a
plane iron quite rapidly.

The color of teak is a light, greyish tan, which will
bleach out to almost white under repeated scrubbings
with salt water. Teak decks generally have the seams
filled with black seam compound for contrast, a traditional treatment that is handsome to the eye of every
sailor. It is considered a sacrilege to paint or varnish over
teak, which is as it should be, for neither is necessary.

Besides decking, cabin soles and cockpit flooring, teak
is used for rail caps, hatches, gratings, coamings, pin-
rails, and exterior trim generally, with no fear of its
checking, splitting, warping or rotting. Truly a magnifi-
cent wood.

A few more words on buying boat lumber are in order. All wood should be *selected,* by you yourself. See what you are getting and select it with each specific job in mind. If you need a plank 6 inches wide and 12 feet long, buy one 14 feet long and more than 6 inches wide. Planks out of stock invariably have checks or splits at one or both ends, which must be cut off. Allow enough extra in width for fitting, or to bypass possible imperfections.

When hunting for oak for knees or stems where natural-grown "crooks" are required, take along cardboard patterns or templates, and don't rely on your eye. You can lay the pattern on a timber and see at a glance whether the grain has the proper curve or not. Look for uniform color—streaks of light canary yellow in white oak often mean diseased wood, and if present will be considerably softer than the surrounding wood. Wood that is to be steam-bent must have straight grain, free of knots. Above all, don't be stingy. By that I mean allow a little extra for fitting, wastage and spoilage. It is surprising how many short-ends, scraps and small pieces are needed in boat construction.

PLYWOOD

There are many different kinds of plywood on the market today, both domestic and imported, some of which are designed specifically for boat construction. Choosing the wrong type can lead to nothing but waste of time and money. A panel of beautifully figured mahogany may be perfect for the enrichment of your living room, but the nature of its inner veneers or core may make it worthless for boat work. There are rigid requirements to be met— it must be waterproof or it will delaminate. It must be rot-resistant, and able to take and hold a variety of finishes without cracking or checking.

As many an amateur has learned to his dismay, common exterior grade fir plywood is not suitable for general boat use. To begin with, manufacturing specifications permit face patches, knotholes, and voids in the core, which allows water to enter the interior of the panel. Rot is then almost inevitable.

Furthermore, the nature of fir wood is such that it will not take a decent finish of any kind. The wood is soft and splintery, and the grain coarse and open. No amount of sealer or paint will prevent its checking. The checking progresses with exposure to the elements, then water gets in and the rot follows.

True marine plywood of mahogany is the only type recommended. It has a minimum of core voids, the plies or veneers are of equal thickness bonded with phenolic-formaldehyde resorcinol glue, which is waterproof, and both faces are of equal quality. It can be had in either rotary or ribbon cut, and takes a fine varnish or paint finish.

U.S. Plywood makes a type called "Weldwood Royal Marine Duraply, which is excellent for planking, where a fine painted finish is desired. It has a tough, smooth face of phenolic impregnated cellulose, somewhat similar to hardboard or masonite in appearance, applied over Philippine mahogany. It needs no sanding or priming, and takes a beautiful paint finish with no grain showing through.

There is a considerable amount of imported marine plywood being used in this country, primarily in the construction of small one-design sailboats. Some makes are of superior quality, with equal faces of choice mahogany, and core veneers of the same quality and thickness. It should be noted that since the metric system is used abroad, what appears to be ¼-inch plywood will be somewhat less—6 milimeters thick, to be exact, and may be 5-ply instead of 3. Where weight is a great factor, as in some of the small one-design classes, a boat built with 5-ply will weigh more than one of 3-ply, and her performance could conceivably suffer thereby.

There are some makes of imported plywood that can only be classed as junk. It is easily recognized by examining the edges—the core will be a single thickness of "gaboon," a cheap, punky white wood. It is completely lacking in strength, and highly susceptible to rot. So consider yourself forewarned!

Some hints on the use of plywood

Working with plywood presents no real problem, but good joinery requires the employment of certain techniques peculiar to the material.

(1) Because of its construction, with the grain of the veneers running in two directions, it must be cut and dressed with special care, or it will tear and splinter easily. To cut it cleanly with power tools, you must use fine-

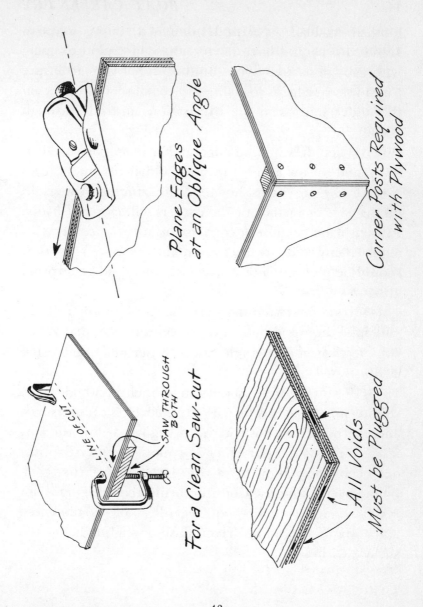

Plane Edges at an Oblique Angle

Corner Posts Required with Plywood

For a Clean Saw-cut

LINE OF CUT

SAW THROUGH BOTH

All Voids Must be Plugged

43

toothed saw blades specifically designed for this purpose. Hand saws should have no less than 11 teeth per inch, with enough set to prevent binding. One trick to insure a clean sharp edge is to clamp a piece of scrap lumber on the under side, as in the illustration, and saw through both.

(2) When dressing the edge with a plane, be sure it is sharp and set for a fine cut, and hold it obliquely, thus making a slicing cut, as shown. *Never* run the plane right off the edge of a panel, or the corners will chip out badly. Always plane from *both* corners *towards the center.*

(3) If any voids are discovered in the core veneers, they should be plugged with a sliver of wood and waterproof glue.

(4) Never put fastenings in the edges of plywood. They will split the wood and have no holding power. When two pieces meet at an angle, use a corner block and fasten as illustrated.

(5) If you are using screws in ¼-inch plywood, don't countersink. The screws will cut their own countersinks, and can be driven flush with, or slightly below the surface. With ⅜-inch or thicker panels, countersink just deep enough to allow for a thin layer of surfacing compound over the heads. Counterboring and bunging is not recommended for plywood, regardless of its thickness. Anchorfast nails can be driven flush or slightly below the surface with a nailset.

Scarfing plywood

While plywood is manufactured in various lengths, many retailers stock only the 4 by 8 size. Now if you want to build a 9-foot dinghy with it, you can't stretch the panels, but you can add on a piece with a scarf joint. It calls for fine craftsmanship and precision, but is not too difficult.

Clamp the panel to a ¾ or 1-inch plank having a sharp, straight edge. Both edges must be exactly aligned. Draw a sharp, fine pencil line 2 inches from the edge, if ¼-inch plywood, 3 inches if ⅜-inch thick, or 4 inches if ½-inch thick.

Now, with a block plane set for a fine cut, start planing off obliquely, a little at a time. The glue lines of the core veneers when exposed will be a guide for even cutting. Don't plane right to the pencil line and to a feather edge, rather, plane until the scarf is *almost* finished. Then unclamp and set it aside.

The lengthening-piece of plywood should, of course, have a grain matching the first piece, of the required length. Allow for the lap and something extra for fitting. Clamp it to the base board, mark and cut as before. You now have a *nearly-finished* scarf cut on each piece, identically.

To finish to a perfectly matched joint, clamp one piece on top of the other, as shown in the illustration, with perfect alignment. Using a fine or medium cut paper, run your orbital sander *very carefully* back and forth over both pieces a few times, until the pencil line is barely visible, and the edge of the lower piece is a mere feather. This should result in a perfectly fitting scarf joint.

Before unclamping, prepare for glueing up, which should only be done under ideal conditions—a dry room with a temperature of 70 degrees. I recommend using resorcinol glue, unless the wood is to be finished bright, in which case I would stick to the urea-resin type, or a clear epoxy glue. Resorcinol glue stains the wood, which is its only undesirable characteristic.

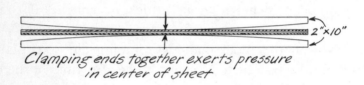

Clamping ends together exerts pressure in center of sheet

The scarf joint must be clamped between two heavy cross-pieces, with even pressure throughout. I suggest two straight 2 by 6 pieces, planed to a slight lengthwise rocker. These, when clamped at each end, will exert pressure in the center as well as the ends, as the illustration shows. You'll also need some sheets of waxed paper, cellophane, or plastic type food-wrap, to place over and below the scarf joint. This is to prevent sticking to the clamping-pieces.

Place the lower panel in position, with the scarf centered on one of the clamp pieces, and secure with a couple of fine brads, clear of the joint, to keep the panel from shifting. Now brush a thin, even coat of glue on both

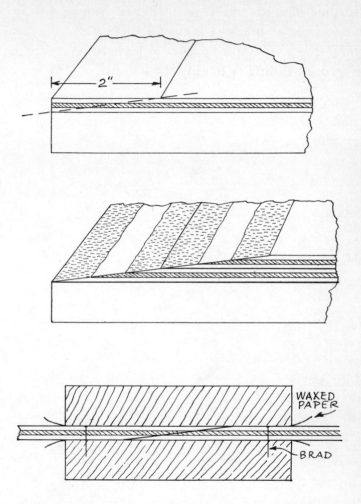

scarfs, place the upper panel in position carefully, with the feather edge just meeting the pencil line, and secure with a couple of brads. Clamp all together, use a heat lamp for quick setting, and allow at least 10 hours curing time before bending or flexing the panel.

III

Fastenings

The amateur who tackles boat work for the first time seldom has more than a superficial knowledge of fastenings, and doesn't realize that the subject involves more than simply driving a nail or screw. But when you consider that the fastenings are all that hold a boat together it becomes a matter of critical importance. If you are to work intelligently and correctly there is much you should know, and I propose to discuss fastenings in considerable detail, with all the facts I consider pertinent, and in accordance with standard boat building practices.

Boat construction involves the use of screws, nails, bolts, drifts and rivets. *Every one of them* must be bored for and installed in a particular manner. To a limited degree they are interchangeable, in that a bolt may sometimes serve in place of a screw, or a drift might double for a bolt under certain conditions. But as a general rule each type of fastening has its specific uses which determine its suitability for the particular job in hand.

There is one vitally important fact to be fixed in your mind right at the start—the cheapest fastening is all too often the most expensive. It is a common fault of all amateurs to put cost as the first consideration. Actually, the fastenings represent such a small part of the total cost of a boat that the difference between the best and the poorest can never justify the sacrifice in quality such an

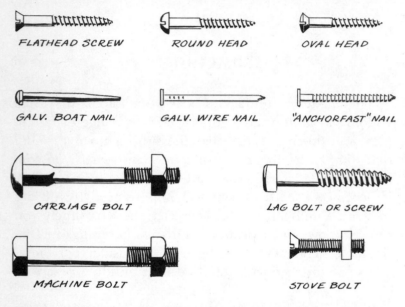

FLATHEAD SCREW   ROUND HEAD   OVAL HEAD

GALV. BOAT NAIL   GALV. WIRE NAIL   "ANCHORFAST" NAIL

CARRIAGE BOLT   LAG BOLT OR SCREW

MACHINE BOLT   STOVE BOLT

Common Boat Fastenings

attitude entails. There are many places in boat building where you can cut costs without compromise, but fastenings do not belong in this category!

Screws

There are more screws used in boat construction than any other type of fastening. Properly used, they have excellent holding power, will withstand severe conditions of service without loosening, yet may be withdrawn without injury to the surrounding wood. They are made with four types of head—round, oval, Phillips and flat.

Round head screws have a very limited use in boats. They are employed primarily where fastenings are temporary, or where they must occasionally be removed. An access panel or section of trim which has to be taken off for the convenience of inspection is a typical example. This type of screw has two bad features. The shape of the head makes it very easy to damage the slot. Unless the screw driver fits the slot perfectly, and is held in exact alignment, it will slip and chew up the head. If this occurs when the screw is half driven there's nothing you can do except withdraw it. Furthermore, in refinishing a boat you can't sand the surface where there are round head screws unless they are first removed. The conclusion is that they are not very desirable.

Oval head screws have the same use as the round head, but they are far superior. The head is much stronger and the slot is not so subject to damage. They should be used with the *countersunk washers,* as shown in the illustration. This makes a neat, professional looking fastening, and is more suitable for frequent withdrawal of the screw. These washers should be brass, and are generally chromium plated. Naturally, the screws should be chromium plated to match. See page 57.

Phillips head screws serve the same purpose as the round or oval type, but are an improvement over both. The X-shaped slot requires a special screw driver. This type of screw is becoming increasingly popular, because it has several advantages not found in any other. First, there is very little danger of the head twisting off or breaking apart when it is driven. Also with this type of slot you are less apt to get a burr, which on an exposed screw head can give you a nasty cut. But most important of all, the slot is self-centering—which means that the screw driver is in perfect alignment with the screw, and, providing you use one of the *correct size,* it will not slip. The conclusion: where it is necessary that the screw head be exposed, the Phillips head is the one to use, with or without a countersunk washer.

Flat head screws are the ones most commonly used in boat work. They must be bored for, and the head is countersunk below the surface of the wood, then plugged, or "bunged." There are four operations involved in boring for the screw correctly. First the counterbore, or hole to admit the screw head; then the countersink, or "chamfered" hole to fit the underside of the head; then the shank or body hole, and finally the pilot or lead hole for the threaded part. Now if you had to use four separate tools for each screw it would be an exceedingly involved operation. However there are tools available which simplify the work.

I have illustrated here a *patented counterbore,* a tool which has been in use by boatbuilders for many years. It performs two operations only—it drills a lead hole and bores for the plug or bung. One set-screw controls the depth of the lead hole, and the other the depth of the counterbore. Thus the lead hole must also serve for the shank hole, and there is no countersink for the under side of the screw head. It comes in only a few standard sizes. Designed primarily for planking fastenings, where

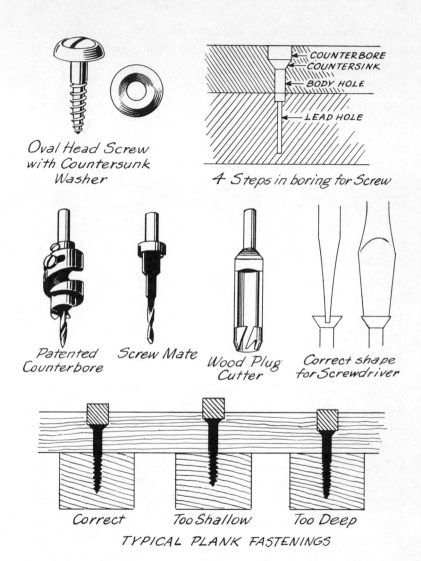

Oval Head Screw
with Countersunk
Washer

4 Steps in boring for Screw

COUNTERBORE
COUNTERSINK
BODY HOLE
LEAD HOLE

Patented
Counterbore

Screw Mate

Wood Plug
Cutter

Correct shape
for Screwdriver

Correct

Too Shallow

Too Deep

TYPICAL PLANK FASTENINGS

screw sizes are more or less standardized, it has a #36 drill for a ⅜-in. bung, and a #30 for a ½-in. bung.

Next I have shown a comparatively new tool, the *Screw Mate* counterbore. This drills the lead hole, the shank hole, the countersink and the bung hole, all in one operation. It is made in the following sizes—

Screw Size	Plug Size
1 " x #8	⅜" x ⅜"
1 " x #10	⅜" x ⅜"
1¼" x #8	⅜" x ⅜"
1¼" x #10	⅜" x ⅜"
1½" x #10	⅜" x ⅜"
1½" x #12	½" x ½"

At first glance this might seem to be an ideal tool, of limited use. However I would point out that it has several faults. There is no stop to prevent boring too deep a hole. Also, it can't be adjusted for different lengths of screws. Finally, it can't be sharpened without reducing the size of the hole it bores. A tool of this design would dull rapidly.

In counterboring for the bung many amateurs (and some professionals) make the mistake of boring too deeply. This results in a weak fastening—for it leaves too little wood under the head of the screw. An old rule-of-thumb is the depth of the bung should be no more than 2/3 its diameter. Thus, ¼ in. is deep enough for a ⅜ bung. The illustration shows a typical situation—a 1-in. plank on an 1½-in. frame, with a #14 by 1¾" screw. Note that the ½-in. bung is 5/16 in. deep. Notice also that the threads of the screw are *entirely* within the frame, thus the maximum holding power of the fastening is utilized.

Bunging

While wood bungs or plugs are available ready-made, of oak, teak, mahogany and pine, you can easily make your own, from scrap pieces of whatever wood you are working with. For this reason I have shown a *wood plug cutter,* a tool which can make a plug any length you desire. It has a ¼-in. round shank for use in an electric drill, hand drill or drill press.

Now let us suppose your screws are all driven home and ready to be bunged. The bungs *must* be made of the same wood the screws are in, and preferably from the same planks—which is a good reason for making your own from the scraps. This is especially important if you are working with mahogany that is to be varnished. There are many varieties of mahogany, and if you use ready-made mahogany bungs there is no guarantee that they will match the color or grain of the plank when finished.

The bungs must be set in some sort of adhesive or compound to keep them in place and to make them watertight. White lead, bedding compound, waterproof marine glue or varnish may be used. If the wood is to be finished bright, I would prefer varnish, because all the others will stain the surrounding wood. For a painted finish the white lead or bedding compound should be slightly thinned. The marine glue is most favored by the professional boat builder because of its superior strength and permanence.

Having dipped the bung well into the varnish or glue insert it in the hole *with the grain running in the same direction as the grain of the plank.* Then tap it *lightly* home with a wooden mallet. Don't use a hammer—you'll either upset the bung or crush it, or both. Lightly does it.

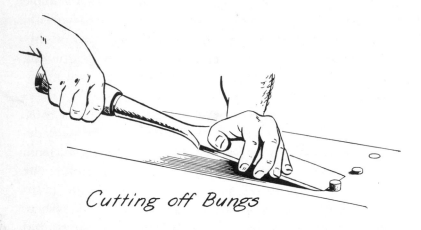

Cutting off Bungs

Before the surface can be finished the protruding bungs must be cut off. At least 24 hours should elapse before starting, to give the glue time to cure. The bungs should be sliced off with a large chisel—the bigger and heavier it is the better. Old timers use what is known as a *slick,* an overgrown chisel 3 or 4 inches wide and about 2 feet long. The weight of the tool gives you better control and a truer cut. The illustration shows how the first cut is made. Most amateurs make the first cut too close to the surface of the plank, and if the grain of the bung happens to run downward the bung chips out below the surface of the plank. Always make your first cut at least ⅛-in. above the surface, then shave it down with successive slices.

Screw drivers

Most amateurs look upon a screw driver as a minor, unimportant tool that you buy in the 5 and 10 cent store. I have observed that few men can resist buying these cheap, poorly tempered, soft steel affairs, particularly if they have a flashy colored handle. They come in mighty handy when prying the lids off of paint cans, or opening stuck windows, but that is about all they are good for.

Buy from a reputable hardware dealer who carries nationally advertised tools, such as Stanley, Millers Falls, etc., and you'll get a screw driver that is correctly designed, precision ground, of carefully tempered tool steel.

The blade of the screw driver must fit the slot of the screw exactly, both in width and in thickness. This means that you don't use one screw driver for all sizes of screws—you'll need several. The tip of the blade should be square, straight and sharp-edged. The two faces should be nearly parallel where it bears in the slot. If the blade is wider than the slot, the edges should be ground down to size. As it wears with use the tool should be touched up occasionally on a small grindstone.

The illustration shows the ideal blade shape. A wedge-shaped, rounded or worn blade is not only useless, but dangerous. Such a screw driver will always slip, ruining the slot and marring the surrounding wood.

Hand screw drivers are generally used for the smaller size screws—say #8 and under. For the larger sizes, which

are most commonly used in boat construction, screw
driver bits and a brace are employed. These bits come in
sets of four from ¼ to ½-in. wide. The brace gives you
the necessary leverage to drive the screw home smoothly
and accurately, and reduces the tendency to wobble and
slip.

A good worker always lubricates the threads of a screw
before driving. The best lubricant I know of is the old
fashioned yellow laundry soap. Drag the screw across
a cake of it and the threads will pick up just the right
amount. One other point to remember is that once you
start a screw in, *don't stop.* Drive it home with one con-
tinual motion. A heavy screw being driven into a hard
wood will invariably "freeze" the instant you stop turning it. I broke many a screw trying to set up just one
more turn, before I learned this little trick.

Composition of screws

Wood screws are made in a variety of metals, each having its own peculiar characteristics which determine its
suitability for boat work. The first requirements are
durability and resistance to corrosion, and strength is
secondary, strange as it may seem.

Galvanized steel screws might, at first glance, seem
appropriate for boat construction, but experience has
proved them to be impractical. They are galvanized after
they are fabricated, and in the process the threads are
filled up and roughened. This makes them extremely
difficult to drive, even when well lubricated. They literally tear their way into the wood, with the result that
their holding power is greatly reduced.

Cadmium plated steel screws are worthless for boat work, for they rust badly. In the same category are steel screws zinc-coated by the electroplating process.

Brass screws are much favored by amateurs, on the weak theory that they can't rust. Well, maybe they can't, but they sure can corrode in a hurry! Common brass is an alloy of copper and zinc, in varying proportions. It is relatively low in strength and hardness. When driving a brass screw into a hard wood, such as well-seasoned white oak, it will very often twist off in the wood, or the head will split off. Because the metal is so soft the slot is easily burred or otherwise damaged.

When brass screws are used in boats that are in salt water, they deteriorate rapidly by corrosion. In a relatively short time the metal becomes dezincified—that is, the zinc has completely leached away or dissipated leaving nothing but porous, crumbly copper. This leaves the screw so utterly devoid of strength that often it is impossible to withdraw such a screw in one piece—it just crumbles to pieces. This is especially noticeable where the screw is in white oak.

I have observed this phenomenon over a period of many years, and have always been curious as to the cause of it. As of this moment I have yet to obtain a qualified answer. My best guess, and that's all it is, is that tannic acid in the wood may cause the dezincification. The high moisture-content of the wood plus salt may also have something to do with it.

On the basis of the above notes it is my opinion that brass screws are definitely unsuitable for boat work. It might be argued that they may be used in the minor, less-

critical places, such as interior trim, with more or less
safety. Well, maybe so, but where do you draw the line?
I would rather play it safe and stick to a metal that is far
superior, which is *Everdur*.

Everdur is recognized by leading boat designers and
builders as an ideal material for screw fastenings, and as
such is invariably specified for yachts of the highest quality. Composed of 96% copper and 4% silicon, it has a
high strength factor and excellent resistance to corrosion.
It is tough enough to withstand severe driving conditions
with no danger of the screw twisting off, which is more
than you can say for brass.

The chart shows the available sizes and lengths of
common wood screws, with the ideal lead and shank
hole sizes for each size screw.

What size screw ?

Unfortunately there is no sure-fire rule or formula by
which you can determine the proper size screw for every
use. It's one of those things you learn by experience,
observation, and the exercise of good judgment and
common sense.

The determining factors in choosing the correct screw
size are 1, the density or hardness of the wood, 2, the
thickness of the member, and 3, the nature and extent
of the stress to be imposed upon it. In plank fastenings
it is customary to use a slightly larger screw in a soft
wood, such as white cedar, than you would in a hard
mahogany, because the larger screw head would be less
likely to pull through the wood. A thin plank requires
a proportionately larger screw than one twice as thick,

TWIST BITS FOR PILOT & SHANK HOLES			Diameter Body	Shank Size	LENGTH IN INCHES																
Hardwood	Softwood	Shank			¼	⅜	½	⅝	¾	⅞	1	1¼	1½	1¾	2	2¼	2½	2¾	3	3½	4
1/32	1/64	1/16	.060	0	●																
1/32	1/32	5/64	.073	1	●																
3/64	1/32	3/32	.086	2	●	●	●														
1/16	3/64	7/64	.099	3	●	●	●	●													
1/16	3/64	7/64	.112	4		●	●	●	●												
5/64	1/16	1/8	.125	5		●	●	●	●												
5/64	1/16	9/64	.138	6		●	●	●	●												
3/32	1/16	5/32	.151	7		●	●	●	●	●	●										
3/32	5/64	11/64	.164	8			●	●	●	●	●	●									
7/64	5/64	3/16	.177	9				●	●	●	●	●	●								
7/64	3/32	3/16	.190	10					●	●	●	●	●	●	●	●					
1/8	3/32	13/64	.203	11						●	●	●	●	●	●	●					
1/8	7/64	7/32	.216	12						●	●	●	●	●	●	●	●				
9/64	7/64	1/4	.242	14							●	●	●	●	●	●	●	●			
5/32	9/64	17/64	.268	16								●	●	●	●	●	●	●	●		
3/16	9/64	19/64	.294	18									●	●	●	●	●	●	●	●	●
13/64	11/64	21/64	.320	20										●	●	●	●	●	●	●	●
7/32	3/16	3/8	.372	24															●	●	●

61

for the same reason. Then again, plywood is never counterbored and bunged, so it too requires a larger screw and its larger head.

Understand that in speaking of *size* I mean the gauge, or body thickness—not the *length*. A simple aid in determining the length is to remember that all of the threaded part of the screw should be *entirely within* the timber to which the plank is fastened.

Plank Thickness	¼"	⅜"	½"	⅝"	¾"	⅞"	1"	1⅛"	1¼"
Hardwood	#6	#7	#8	#9	#10	#12	#14	#16	#18
Softwood or Plywood	#7	#8	#9	#10	#12	#14	#16	#18	#20

The above chart is shown merely to demonstrate one application of screw sizes—the planking of a hull. It will be of little help in the framing of a boat, nor is it an inflexible rule for planking. The sizes specified are only average and not at all critical—a #12 might be substituted for a #10 without causing a major catastrophe. But at least it provides a starting point for your thinking, and should prove a help in learning to use screw fastenings properly.

Nails

There are four types of nails used in boat work. Actually there is a fifth—the common copper wire nail, but it is used only as a rivet, so does not properly belong in this category. First, there is the galvanized boat nail, used primarily in planking. Then there is the common wire nail, also galvanized, used in framing but rarely in planking. Galvanized finishing nails or brads are useful in interior joinery.

Finally, there are the barbed or serrated wire nails of copper, bronze or Monel, generally called "Anchorfast." Actually, "Anchorfast" is the registered trade name of International Nickel, and is properly identified only with monel.

Monel Anchorfast boat nails are far superior to common wire nails, galvanized boat nails or screws. The shank of this nail is barbed, with backward-pointing annular serrations that grip the wood fibers with tremendous holding power. For a *permanent* fastening, as in planking, it is superior, preferred by many to screws. Once driven into hard oak, it can be drawn only with difficulty, tearing the wood as it comes. These boat nails come in lengths from 1-inch to 3-inches in length. Anchorfast common nails are made as small as ⅝-inch, and are admirable for plywood fastening.

The same type of nail is made by various manufacturers of copper and silicon bronze, each having its own trade name. Monel is an alloy of copper and nickel, and while its cost is higher than other non-ferrous metals, its superior qualities more than justifies its use.

There is another factor worth remembering. The use of Monel fastenings means a higher resale value for your boat. No matter what the ravages of time and hard usage may do to her, you can be sure that the condition of those fastenings will remain unchanged, and therefore there will be less depreciation.

When planking with Anchorfast nails you counterbore and bung, just as you do with screws. A lead hole should be drilled for easier driving, but it must be *no greater than ⅔ the diameter of the nail,* otherwise the holding power will be destroyed. Many professionals use a power hammer and drive the nail home without boring a lead hole, which results in a considerable saving in labor costs. But since we are amateurs, it's best to bore for the nail and make the job a bit easier.

The nail should be driven in nearly flush with the planking, then set into the counterbore with a "spikeset" or heavy punch. In working with plywood the operation is a little simpler, since you do not counterbore and bung. With ¼ in. or ⅜ in. plywood it is not necessary to bore a lead hole—just drive the nail home and set the head in flush with a nailset. With the heavier sizes, ½ in. and up, it is best to bore a lead hole and countersink, but the countersink should be *no larger* than the diameter of the nailhead. Then the nail can be set a little below the surface, say ¹⁄₃₂ in. or ¹⁄₁₆ in., and the shallow hole can be filled and smoothed with trowel cement.

SIZES AND WEIGHTS OF MONEL ANCHORFAST BOAT NAILS

Gauge	Wire Diam.	Penny size	Length	Approximate Number per Pound
#14	.083″	4d	1½″	400
#12	.109″	2d	1 ″	350
		3d	1¼″	280
		4d	1½″	230
#10	.134″	4d	1½″	135
		5d	1¾″	120
		6d	2 ″	105
		7d	2¼″	94
		8d	2½″	84
#8	.165″	4d	1½″	90
		5d	1¾″	83
		6d	2 ″	75
		7d	2¼″	64
		8d	2½″	58
		9d	2¾″	53
		10d	3 ″	48

In fastening the butts in planking I would use Everdur screws, even though the rest of the fastenings were Anchorfast nails. In the first place the plank butts generally stand off a bit from the adjacent planks and must be pulled in to the butt blocks in order to be fair. Screws naturally will draw the plank in better than nails. Sec-

SUGGESTED SIZES AND SPACING OF ANCHORFAST NAILS

For Planking and Decking (Lumber)

Plank Thickness	Gauge	Length
½"	#14	1¼"
⅝"	#12 #10	1½" 1¼"
¾"	#10 #8	1¾" 1½"
⅞" or 1"	#8	1¾" or 2"

For Plywood—Planking and Decking

Plank Thickness	Gauge	Length	Spacing on Edges	Spacing for Frames & Beams
¼"	#14	⅞"	1½" to 2 "	3" to 5"
⅜"	#12	1¼"	2½" to 3½"	4" to 6"
½"	#10	1½"	3½" to 4½"	5" to 6"
⅝"	#8	2 "	4"	6"
¾"	#8	2¼"	4"	6"

For Edge-nailing Strip Planking

Strip Size	Gauge	Length	Spacing
⅝"	#14	1½"	4"
¾"	#14	1¾"	4"
⅞"	#12	2 "	5"
1 "	#12	2¼"	5"
1⅛"	#10	2½"	6"

ondly, the butt blocks are but little thicker than the planking, and there just isn't enough wood to grip a nail properly.

In framing, as in planking, where nails or screws are indicated, Monel Anchorfast Nails are superior to other types of fastenings. It is impossible to tell you what gauge and length to use for every instance—it's one of those things you learn from experience. I would like to point out, however, that Monel is a very tough metal, so in choosing the gauge of the nail you should be more concerned with the size of the *head* than the *shank*. The heavier the gauge the larger the head. If you use a gauge that is a little too light, there is more danger of the head pulling through the wood, than there is of weak holding power in the shank.

Bolts

In framing, and in through-fastening most structural members, it is customary to use bolts—hot-dipped galvanized steel, Everdur (bronze), or Monel. Bolts are also required to secure the ballast keel in sailing craft, for fastening hardware and fittings to the hull, and occasionally in planking.

Machine bolts, the heads and nuts of which may be either square or hexagonal, are used primarily for fastening ballast keels, and rarely elsewhere. Their composition depends upon the metal used in the ballast casting. With iron, *only galvanized iron or steel* should be used. Lead keels *must have bronze or Monel bolts.* Any deviation from these combinations means the immediate start of galvanic corrosion, and the eventual loss of the ballast keel.

Stove bolts, of brass or bronze, have a flat head like a wood screw, and are used only for fastening deck fittings and hardware, with possibly one exception. Very often in planking, the butts stand off from the hull so far that it is difficult or impossible to bring them in and hold them to the butt-blocks with screws or nails. The solution is to clamp them in and fasten with stove bolts, counterboring and bunging just as you would with screws or nails. Only bronze or Monel bolts should be used if the boat is for salt water.

Carriage bolts are used in framing and in fastening most structural members. They have a crowned head, and a short section under the head is square to prevent the bolt from turning when the nut is set up. They are obtainable in galvanized steel, bronze and Monel, from $1/4$ in. to $3/4$ in. in diameter, and in lengths up to 20 in. Needless to say, washers should always be used under the nuts to secure maximum bearing. Don't depend on the nut to draw the square section under the head into the wood—drive the bolt home with a hammer until it's well seated. In using the galvanized bolts it is a good idea to coat the shank with red lead before insertion, and to red-lead the nut and bolt end after setting up, as a rust preventive.

Lag bolts, sometimes called lag screws, really should be classified as screws, since they are nothing more than heavy wood screws with square bolt heads. They run from $1/4$ in. to $3/4$ in. in diameter, and are available in galvanized iron or steel and bronze. Bronze lags are generally used to fasten stuffing boxes and stern bearings, and galvanized iron lags are occasionally used to secure the motor to the engine beds.

Lags are generally disliked, mainly because extreme care must be used to set them properly, and unless they are bored for accurately their holding power is very uncertain. The length of the shank from the underside of the head to the first thread must be measured, and a hole exactly that depth must be bored, the same diameter as the shank. Then a lead hole about seven-eighths the length of the threaded part is bored, with a diameter no greater than two-thirds of the shank. The lag should be well lubricated with soft soap or grease before driving.

Drift Bolts

This form of fastening is simply a rod of galvanized wrought iron or steel cut to the required length and driven into a hole bored slightly less than the diameter of the rod. Drift bolts are used for fastenings in keel timbers, deadwoods and all heavy members in the keel assembly of large vessels. They are also used for edge-bolting rudders, centerboards and planked transoms. Log rails are edge-fastened to the clamps or sheer strakes with drifts. Cabin trunks are often edge-bolted to the carlins in a similar manner.

Few amateurs know how to use or install drift bolts correctly. Considerable skill and ingenuity is required, both in boring and in driving, and unless you have a thorough understanding of all the techniques involved you will soon have a lap-full of trouble. In view of this, I propose to discuss the subject in considerable detail.

First, the term "drift bolt" may mean one of several things. In general, it is a *long* fastening. When driven only part way through the timbers, as in the first illustration, it is essentially a nail or spike. When it is a through-fastening, as in the second illustration, it is in effect a bolt. All through-fastening drifts are driven with a *clench-ring* under the head, which should be of the same metal and of a size to fit the drift. A clench-ring is a form of washer, flat on the bottom with the top crowned and countersunk for the drift bolt head. The impact of the sledge or hammer used in driving will generally upset the top of the rod sufficiently to form a head with a proper shoulder to fit the countersink. If not, it should be shaped by tapping with a ball-pein hammer. The point of the drift is generally riveted or peined over another clench-ring, except where the drift is used to through-fasten a plank on edge, such as a cabin trunk. In this case the point has a thread run on it and the drift is set up with a nut and washer. See page 79.

Methods of boring for drifts vary with the builder. Ship augers are most commonly used, although some prefer long-shank twist drills. Ship augers run from 14 to 20 inches in length, and should be of the single-cutter, barefaced type. By barefaced we mean that it has no screw or worm. If the depth of the hole to be bored is greater than the length of the auger, the shank of the auger should be welded to a length of drill rod of the same diameter. It is then cut to the desired length, and can be used in a brace or an electric drill. Twist drills must be welded to a length of drill rod in a like manner.

As I have said before, a great deal of improvising is

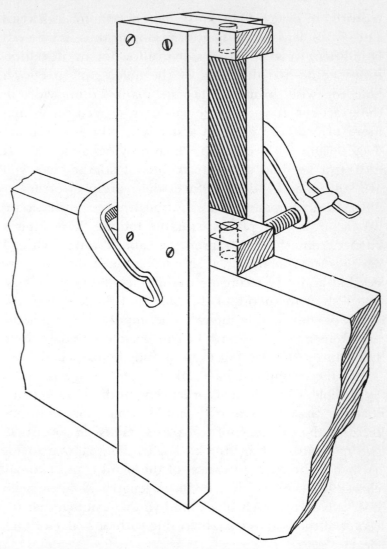

Jig for boring holes for drifts

required in boat building. To bore a hole for a drift in a heavy timber is a relatively easy job, for there is generally sufficient wood on either side to allow for any deviation from the desired direction of the hole. But to bore a hole edgewise through a 1½ in. plank 17 in. wide, in the exact center without deviation, you need something more than a keen eye and a steady hand. For accurate deep boring you need a jig, which provides a mechanical means of keeping the auger centered all the time you are boring, and greatly reduces the margin of error.

In the accompanying illustration I have shown a typical jig for boring planks, suitable for use in making a rudder, centerboard, transom or cabin trunk. While I have shown dimensions, it should be noted that they are flexible, and vary according to the thickness of the plank, its width or the depth of the hole, and the length of the shank of your drill or auger. The important thing is the principle involved, and you can change or adapt it to suit your particular job as your judgment dictates.

For the example, I have taken a 1½ in. plank and a ½ in. hole, which is in accordance with the accepted rule that the diameter of a drift should be no more than one third of the thickness of the plank. This jig consists of two oak blocks, 1 in. thick, 3½ in. long, 1½ in. wide (which is the exact thickness of the plank), and two oak cleats, ¾ in. by 2½ in., 16 in. in length. A ½ in. pilot hole is bored in each block, and you will notice that the center of the hole is exactly in line with one edge of each cleat.

To use the jig you first strike a pencil line across the edge of your plank to mark the center of the drift, then square the line across the face to mark the line of travel. Now slip the jig over the plank like a clothes-pin, line up the edge of the cleat with the pencil line and clamp it in position with a large C-clamp. Set the clamp up tightly so there's no possibility of the jig shifting. Set your auger into the pilot holes and you are ready to bore.

If this jig is to be of any value it must be made very carefully and accurately. Lay out the center lines for the pilot holes in the blocks, 1 in. from one end, and square the lines down over the edges. Bore the 1/2 in. holes in a drill press so that they will be exactly centered. The cleats can be fastened to the blocks with #8 wood screws, 1 1/4 in. long.

In boring for drifts, or all deep boring, amateurs invariably try to push too fast, and the bit gets clogged with chips, generally starts to "run" one side or the other, and sometimes breaks off in the wood. A good workman continually runs the bit in and out, actually cutting but a few revolutions at a time, and exerting no more pressure than is necessary. This clears the bit, keeps the chips moving, and reduces the possibility of the bit running off at a tangent.

All drift bolts, where possible, should be canted or raked slightly, which greatly increases their holding power. The boring jig is used here also, canted to the desired rake. I have shown a typical centerboard layout, with the drifts plotted. Notice that there are *through-fastening* drifts in each end of the board. Now this involves edge-boring through all five planks—a long straight hole with no deviation whatsoever, else the long drift could not be driven.

The proper procedure here is to lay the tapered boards together, in position, on a couple of horses, and clamp them so they can't shift. Now with a straight-edge mark the line of the two long end drifts across the boards with

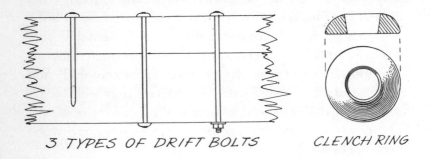

3 TYPES OF DRIFT BOLTS   *CLENCH RING*

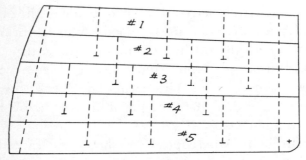

TYPICAL LAYOUT OF DRIFT BOLTS
IN CENTERBOARD

a clean pencil line. Mark also the line of the rest of the drifts indicating by a mark their exact length or depth. Also number each board to indicate its position. Now remove the clamps and pick up the top board, and set it up for boring. Clamp on your jig, with the cleat always on the left side of the line, and bore each hole in the plank. Next set up the second plank and bore, but note that only the end holes go clear through the board— the others must be measured for length and bored accordingly. This means some form of a stop must be used on the auger to prevent boring too deep. The simplest stop consists of a length of adhesive tape wrapped around the shank, as a marker.

In assembling the centerboard you work from the bottom up, driving first the drifts through #4 into #5, then #3 into #4, etc. The through-fastening end drifts are put in last.

After a drift has been cut to length the point should be tapered. Lay it flat on an anvil or block of steel and hammer the end to a slightly rounded, blunt point. You don't need a long taper—just strike the last half-inch or so. This blunt point prevents the end from tearing the wood and makes it easier to drive.

Lubricate the drift with soft soap and drive it with a maul or heavy hammer. The fewer the number of blows required to drive it, the easier it will be. Using too light a hammer and too many blows will cause the head of the drift to mushroom badly. This is even more serious if the drift is bronze, since bronze is much more malleable than wrought iron or steel.

Rivets

Copper rivets are the standard plank fastenings for lap-strake construction. Elsewhere their use is restricted primarily to very light, thin planked boats such as dinghies. A riveted, lap-strake hull is very flexible, and ex-

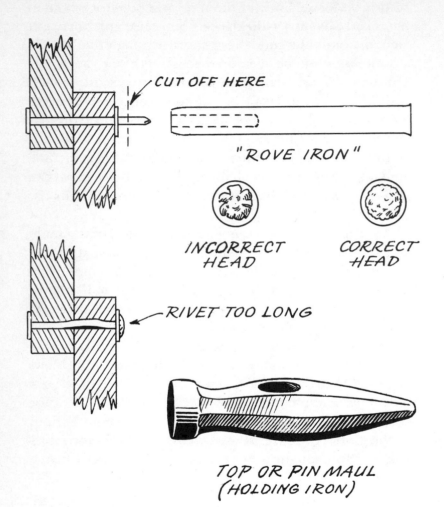

CUT OFF HERE

"ROVE IRON"

INCORRECT
HEAD

CORRECT
HEAD

RIVET TOO LONG

TOP OR PIN MAUL
(HOLDING IRON)

ceedingly strong, but it calls for a higher degree of craftsmanship than any other type. Since the plank seams have no caulking, you must know how to do a professional riveting job or a leaky hull will result.

This type of fastening is a copper wire nail riveted over a small copper washer called a "burr", or "rove." When you buy the copper nails and burrs to fit, you will discover that the burr cannot be slipped on the nail with the fingers alone—it must be driven on with a tool. This always surprises most amateurs, for the natural supposition is that they have been given the wrong size burrs. The tool used to drive the burr on the nail is called a "rove-iron," and is nothing more than a hollow punch. Since it is not an easily obtained tool, I have shown a sketch of one with the suggestion that you make your own. Take a piece of iron rod $\frac{3}{8}$ to $\frac{1}{2}$ in. in diameter and about 4 inches in length and drill a hole in one end. Make the hole 1/32 in. greater than the diameter of the copper nail you are using, and about 1 inch in depth.

Every rivet must be bored for, and the size of the drill must be no greater than the diameter of the nail. In fact, if the planking is a relatively soft wood the hole should be slightly smaller.

After the hole is bored the nail is driven home, with the head well seated—countersinking is not required. Then place the burr on the point and drive it on with the hollow punch till it seats well against the wood. The next step is to cut the nail off with a pair of nippers, leaving just enough metal to form a good head over the burr. Here is where an amateur often goes wrong. If the nail is cut off too close to the burr, the head will be so thin after riveting that it will have very little holding power. If too much of the nail is left, the head will be upset or cocked, and the nail will bend somewhere within the wood. Note that the sketches show clearly what a poor fastening this would make.

The amount of metal projecting above the burr after the point has been nipped off should be approximately one and a half times the diameter of the shank of the nail. This is but another of those skills that come with experience, for good riveting requires much practice.

The actual riveting is done with the rounded end of a small ball-pein hammer, while a heavy steel or iron weight, called a *holding iron*, is held against the head of the nail. The cut-off end of the nail is shaped to a slightly rounded head by a rapid succession of very light blows or taps. It is a common fault of the inexperienced to use too heavy a hammer and to strike too hard. This results in a smashed, split or mushroomed head, and crystallized metal, as you will see in the illustrations.

Copper becomes hardened when hammered. Thus the rapid, light peining not only forms the oval head, but increases the holding power of the rivet. The peining also draws the rivet up, much as tightening a nut does on a bolt. Actually, you rivet by ear—the initial taps produce a dull sound, and as you progress the sound becomes sharper, higher pitched, and more *solid*. This progressive change of pitch is very noticeable, and you soon learn to recognize the point at which the rivet is properly set up.

Your holding iron should be as heavy as you can

handle conveniently. An old flatiron, a short piece of heavy shafting, or a broken sashweight would do the trick. I prefer a top or pin maul, *without the handle*, weighing not less than 5 pounds, using the head for general use, and the point where there isn't much room to work. In case you do not know what a pin maul is, I have shown a drawing of one.

IV

Marine Glues and Adhesives

Revolutionary changes in boatbuilding methods have taken place in recent years, and the end is not in sight. As in other industries, technological development in the field of chemistry has put new tools in the hands of workers, and we can do things with wood that were once impossible. Preservatives, surface coatings, adhesives, laminates, plastics and synthetics are products of common usage, and enable us to build boats that are stronger, require less maintenance, and last longer.

In addition to the primary adhesives—the resorcinol and epoxy glues—there are numerous flexible sealants and caulking compounds of synthetic rubber, polysulfide, polyurethane and silicone rubber. They supplant the old fashioned caulking cotton, can be used as bedding compounds, and in some cases, are actually adhesives.

They have their limitations as well as their uses. Because we have waterproof glues of tremendous strength, there is no reason for throwing away our nails and screws, nor do synthetic coatings that are seemingly indestructible make paint and varnish obsolete.

In my opinion, the greatest contribution that modern

chemistry has made to boatbuilding industry is in the field of adhesives, specifically the various resin glues and epoxies. Phenolic resin glues, which require both heat and pressure, have made possible our marine plywood and the moulded plywood hull.

To employ these resin glues successfully, and to develop their maximum strength, you must have a clear-cut understanding of their characteristics and limitations, of gluing procedures in general, and of the craftmanship involved. So before we take up the specific glues in detail I would like to discuss these important factors at random, to establish what might be called the proper mental attitude.

To begin with, you must not look upon glue as an antidote for poor workmanship. Too many amateurs let a badly fitted joint pass, in the mistaken belief that the strength of the glue is sufficient to counteract their lack of skill. A good gluing joint should be so fitted that the two surfaces are in perfect contact everywhere— wood-to-wood throughout, and no voids anywhere. Such a joint takes time and skilled workmanship. To be sure, these glues will bridge a gap of moderate size, but their tendency to shrink while curing, plus the lack of penetration at that point, results in uneven strength and a vulnerable spot when stressed.

Another point to remember is that these glues are cured by catalytic action, which means that a chemical change takes place by means of a catalyst, which itself is stable. Depending on conditions, and the nature of the resin, the duration of this change might be anywhere from a couple of hours to nine or ten days. Since this chemical action starts immediately the mixture is prepared, there is a limit to its working time. Temperature, humidity, and the moisture-content of the wood are very important factors.

Now the manufacturers of these glues, naturally, supply adequate instructions for their use. They include mixing directions, working times, temperatures and pressure periods. Naturally you have a certain amount of latitude, and the limitations are clearly stated. But unless you follow the manufacturer's recommendations and keep within the prescribed limitations you are asking for trouble, and you'll have a glue joint of questionable strength. To fully exploit these glues, to obtain their maximum performance, you must strive for as nearly ideal conditions as possible. Where a joint has orthodox fastenings as well as glue, you are in a measure protected from dire disaster. But where you put your trust only in a thin glue line and a prayer, as in a hollow mast, your workmanship had better be good!

Of the urea-formaldehyde resin glues, Weldwood is undoubtedly the best known. It requires no special techniques or skills, is easily handled, and very inexpensive. It is advertised as water-resistant, but within certain limits it is practically waterproof. I would not use it where it would be submerged in water for long periods of time, but is a good all-purpose glue for non-critical areas.

It is prepared for use by mixing with water to the consistency of heavy cream. It hardens very slowly, so you have plenty of working time. As with other glues, 70 degrees is an ideal temperature for working, and curing time is lengthened with lower temperatures.

Another factor that affects both the curing time and the strength is moisture—the moisture in the air and the moisture content of the wood. High humidity can prolong the curing time to the point where the strength of the glue is way below normal. The moisture content of the wood should be no higher than 15% nor lower than 5%. Of course few amateurs can be expected to bake test samples every time they have a glue job, nor is it necessary. It is simply a matter of using common sense—don't try to glue wood that has been submerged or rain-soaked until it has had ample time to become reasonably dry.

Resorcinol glues are derived from a phenolic plastic, and Elmer's Waterproof Glue, a product of the Borden Company, is the best known. It is completely waterproof, much stronger than urea resin glue, and will withstand continuous immersion in cold or boiling water.

It comes in two parts, a dark red, liquid resin and a powder catalyst, and it is prepared for use by mixing 4 parts of liquid with 3 parts of powder, by measure. Measuring must be *accurate,* for if too much liquid is used the glue will not harden or cure correctly, and too much catalyst will cause the glue to harden so fast that there will not be time for handling. For mixing small quantities it is best to buy two sets of kitchen measuring spoons, preferably glass—one set for the liquid and one for the powder. A glass container should be used for mixing. It should be noted that the liquid resin is of a very syrupy nature, so when only a small quantity is needed, allow a little extra to make up for the amount that clings to the spoon.

A temperature of 70 degrees is stipulated by the manufacturers themselves for proper gluing. Now this is easily complied with if you are working in a heated room with a thermostat, but if your boat is out of doors in your backyard, or afloat at her moorings, temperature is a matter of chance. If the temperature goes above 70 degrees there is nothing to worry about—the glue sets up much faster and your assembly time is that much shorter, so you work a little faster. But if it is below 70, say 55 or 60, you either improvise or you don't glue. If you are within reach of household electricity with the aid of an extension cord, a 100-watt light bulb or a heat lamp will bring enough heat to the spot where it is needed. Hang the lamp close to the glued joint immediately after pressure is applied, and drape an old tarpaulin or blanket loosely over the assembly. But if you can't bring 70 degrees to the spot artificially, don't attempt a glue job—wait until it comes naturally.

The strongest of all glues, and the most expensive, are the epoxies, and there are many different types on the market. They vary in consistency, color, strength, flexibility, and in assembly and curing time. Proportions in mixing the resin and catalyst will vary greatly with the different brands and formulations.

One great superiority of epoxy glue is that glued joints do not require pressure, but have merely to be placed in contact until the glue has set. Joints do not have to be exactly fitted as with other glues. In fact, a film of glue should remain in the joint between the two surfaces for best results. Thus, the adjoining surfaces should be *sawn*, rather than planed or sanded.

Many amateurs shy away from epoxy glues because of

their high cost, but in certain critical applications they are almost indispensable. Very hard, dense woods such as teak, and most oaks, are very difficult to glue with any certainty that the joint will hold indefinitely. Epoxy glue is the only adhesive that will assure a permanent bond.

A proper job of gluing can not be done without pressure. Pressure can be had by means of clamps, weights, rope, nails, screws or bolts, depending on what is being glued. Pressure is applied for two reasons, to insure maximum penetration of the glue into the wood, and to force out all air that could otherwise form voids or pockets. Soft woods, such as cedar or white pine, can be glued with light or moderate pressure. Woods of high density, particularly oak, require high pressure for a good bond. When clamps are used, always put bearing blocks under the clamp jaws to avoid crushing and marring the face of the wood being glued.

Now suppose we take up the procedure, step by step, in a typical glue job. Before any glue is mixed or applied we make a "dry run." Bring the two members into position and check for a perfectly fitted joint—wood to wood and no daylight in between. Clamp or wedge the parts together firmly in the desired position. Now if there is any possibility of the two members sliding out of alignment when finally assembled, strike one or more pencil lines across both pieces as check marks. Just line up these marks in the final assembly and you will know that the two members are properly positioned. While still under clamps, bore for the fastenings, if any, and get your fastenings and the necessary tools ready at hand. Check the temperature—if it's a hot day, in the 80s or 90s, remember that your assembly time will be no more than 10 or 15 minutes.

Now remove the clamps and place both members side by side or in a convenient position for spreading the glue. Mix up the glue by careful measure, adding the catalyst to the liquid resin and stirring until it is free from lumps. No one can tell you how much glue you'll need for a given joint, but it's better to mix a little more than you need than too little. Apply the glue to *both* surfaces, evenly, with a brush or knife. If you are gluing soft woods you can assemble the parts immediately, but if you are using oak let them stand for 5 to 10 minutes to allow the glue to penetrate into the wood.

Bring the parts together quickly and check for position. Put in the fastenings, if any, and then apply your clamps. If sufficient glue has been spread and the proper pressure applied, a thin bead of glue will be squeezed out of the joint. Once the clamps have been set up, spaced so that even pressure is applied, it should be undisturbed for not less than 8 to 10 hours. The higher the temperature the shorter the pressure period. Although the glue may be dry after 10 hours under the clamps, the chemical change continues, and maximum strength is not attained in less than 6 days, at 70 degrees average temperature.

Because of the extremely high sheer strength and waterproof characteristics of the glue, there is practically no end to the possibilities in its use, whether it be framing, planking or interior cabinet work. One important application is in laminated members. When a knee is required, the amateur invariably has difficulty in locating natural-growth material of the right curvature and shape. I have illustrated a laminated knee which is easily made to fit your own specifications. White oak should be used, the laminations no thicker than can be bent to the required curve without steaming. Naturally, you have to build a strong form over which the strips are bent. The strips

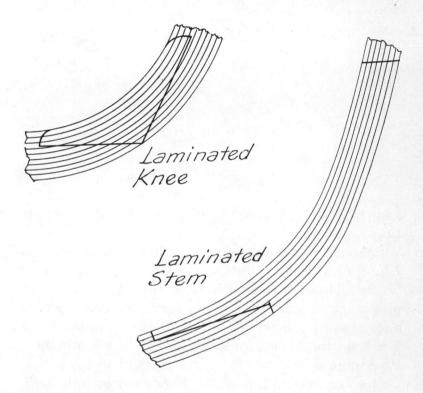

Laminated Knee

Laminated Stem

should be longer than needed, for ease in handling. Spread the glue on each strip in turn and lay them up bread-and-butter fashion, then clamp the whole assembly at once. After curing, the pattern for the knee is laid out

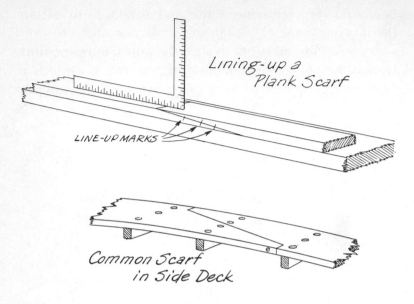

Lining-up a Plank Scarf

LINE-UP MARKS

Common Scarf in Side Deck

and cut to size. A knee of this type is actually *stronger* than a grown one.

Laminated stems can be made in the same manner, as I have illustrated. Another application is in cabin-roof beams. These are normally sawn from a plank to the proper crown, and if the crown is considerable, there is a lot of waste lumber. A built-up beam, laminated and glued, is stronger, size for size, than a sawn one, and there is no waste at all.

Open cockpit skiffs normally have a narrow side deck, one plank wide. Since it is impossible to spring the plank in place, edgewise, it must be made of short lengths cut from wide planks. In order to save time and keep the cost down, most professional builders simply butt the planks, with a butt-block underneath. Invariably these butt joints open up after a few seasons' use, and water gets in to start the inevitable rot.

A far better practice is to scarf the joints, as in the illustration, using waterproof glue. Thus the side deck is essentially in one piece, much stronger, and permanently watertight. Note that the scarf must be centered over a deck beam for properly locating the fastenings.

V

Joinery

THE SCARF JOINT

The scarf is a joint peculiar to ship and boat construction, as indispensable today as it was hundreds of years ago. It is most commonly used in joining the various members of the backbone or keel. Occasionally it is used in planking, and in various members of the frame and deck. Hollow box spars are built up from short-length stock by the use of scarf joints, at a considerable saving in cost with no loss of strength.

In laying out a scarf, irrespective of where it is used, there is one principle requirement to be met. The line of the scarf joint *must not* cross the grain of the wood at an angle greater than 5 degrees, as indicated in the illustration. This 5-degree slope can be expressed in lineal measurement as a 1 inch rise in every 10 inches of run. Anything less than this means a weak joint.

In the accompanying sketches I have shown the three most commonly used scarfs. The first is a plain plank scarf, used as a glued joint, in spar making, and in joining planks in lap-strake hulls without butt-blocks. The

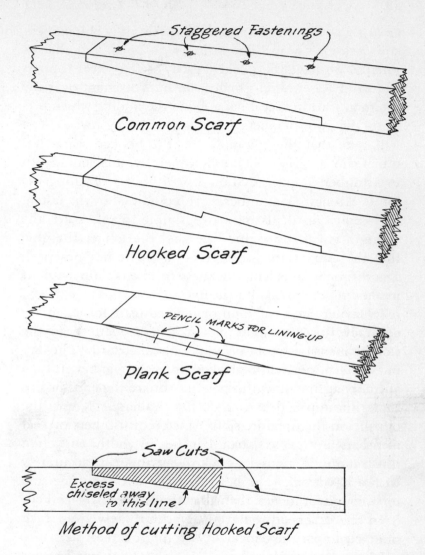

Staggered Fastenings

Common Scarf

Hooked Scarf

PENCIL MARKS FOR LINING-UP

Plank Scarf

Saw Cuts

Excess chiseled away to this line

Method of cutting Hooked Scarf

next is a common scarf, used primarily in keel members, and occasionally in bilge stringers, clamps, and shelves. Finally, we have the hooked scarf, used in keel timbers and rarely elsewhere, which is an improvement over the common scarf in that it has a jog in the middle which prevents the two members from working lengthwise. You will note that the common and the hooked scarfs terminate in a "lip." There is generally sufficient depth in a timber to cut a shallow lip without weakening the joint. Strength in a timber scarf is obtained by the fastenings—bolts or drifts. In the simple plank scarf the strength is had by a glue line, and the longer the glue line the greater the strength, so a lip is not practical. There is not sufficient thickness to cut the lip without weakening the plank drastically.

In laying out a common or hooked scarf for a timber, accuracy is very important, so make a pattern to the exact dimensions on heavy paper or cardboard. The cutout pattern can then be tacked on each member in turn and the outline drawn in pencil. Square the ends or lips across the top and bottom of the timbers and trace the outline on the opposite side. Since most timbers or keel members are not less than 4 inches in width, and often much more, it is quite a trick to cut the scarf accurately with a hand saw. The best way is to cut from one side for a couple of inches, then turn the timber over and cut from the other side. By cutting alternately from each side you lessen the danger of running off the line.

When both timbers have been cut, fit them together and see that they are perfectly straight lengthwise. Now if you have a perfectly fitting joint you should not be able to see daylight through it anywhere. It must be wood-to-wood everywhere. An old trick in fitting a joint is to chalk one member with blue carpenter's chalk, then bring both pieces together and set them up tightly with C-clamps. Then take them apart, and if the joint is perfect the chalk should be evenly transferred to the second member. If the chalk appears only on certain areas, those areas represent high spots which should be cut down with a small block plane or a wide chisel.

The method of cutting a hooked scarf requires some explanation. A little study will show that it can't be cut entirely with a saw. In the second illustration of the hooked scarf you will note that the outer half, or end piece, will be removed by two saw cuts, but the shaded half must be taken off with a chisel. In very heavy timbers, a shipwright would dub this off with an adze—a special narrow adze known as a "strap-adze," made specially for cutting scarfs. The amateur, however, is rarely skilled in the use of an adze, and a heavy chisel will do the trick.

The chisel should be no less than 1½ in. wide, for a job like this, as it is a heavy cutting, and use a wooden mallet—not a hammer. Don't try to take too big a bite; thin chips mean more control, and less danger of splitting off more than is required. Cut both ways towards the "shoulders," and when you near the finish line, discard the mallet and finish by hand. A small block plane may e used where there is room, but the chisel must be elied on to make the shoulders true and square.

In boring for the fastenings in scarfs, see that they are taggered slightly, rather than right on the center line, so that they do not all go through the same line of grain.

Otherwise, there is danger of the timber splitting down the line of the fastenings, and even though it doesn't split it will always be a weak spot—right where you need strength.

In joining the various members of the keel—stem, knees, horn timber and deadwood—it is of the utmost importance that no water be allowed to enter the joints or scarfs and start rot. In olden times the inner surfaces were coated with tar or pitch. Later they were bedded with thick paint or white lead, and eventually various bedding compounds were developed for the same purpose. All of these, however, had a common fault. Over a period of time the oils or other vehicles would either leach away or be absorbed by the wood, and the compound would lose its elasticity and become hard and crumbled. Thus with the bond destroyed water would inevitably get in and rot would start.

This trouble has been practically eliminated with the improved compounds now available, such as the latex or polysulphide polymer types. They retain their elasticity indefinitely, and form a permanent bond to the wood. However, because of their elasticity these rubber compounds add very little to the strength of the joint, and the load is carried primarily by the fastenings. It is for this reason that many professionals prefer a glued joint. The use of an epoxy glue, with its tremendous sheer strength, takes some of the load off the fastenings, adds greatly to the rigidity of the joint, and prevents working of the members. As stated before, you should strive for ideal conditions when using the epoxies—proper temperature, low humidity and dry wood.

The simple plank scarf requires more skill to make

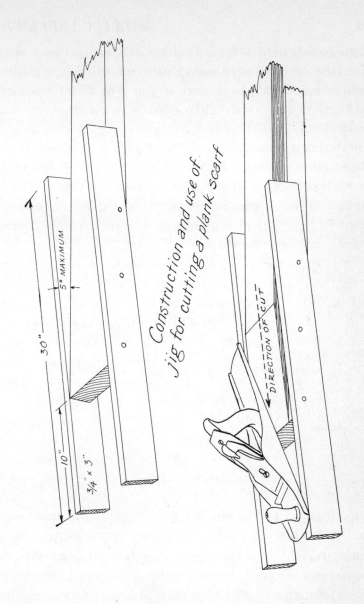

Construction and use of
jig for cutting a plank scarf

30"

5° MAXIMUM

10"

3/4" × 3"

DIRECTION OF CUT

than a timber scarf, for a mechanically perfect joint must
be made if it is to have any value at all. Since its strength
is determined by the quality of the glue joint, both surfaces must be exactly alike—square with the edge and
face of the plank, with the same angle of cut, and in
absolutely perfect contact. If the two planks are not
sliced off at the same angle they will be out of line when
glued up, even though the error be very slight. If one
surface is not square across the face, the finished assembly will have a twist at the scarf, and the two planks will
not lie in the same plane.

The easiest way to make a perfect scarf joint is to use
a jig, such as I have illustrated. It consists of a 2-inch
plank, slightly wider than the stock you wish to scarf,
and two side pieces which serve as guides or runners for a
hand plane. Let's assume that the planks to be scarfed
are 5 inches wide. Select a 2 in. by 6 in. plank, 6 or 8
feet long, that is straight, square, clean and unwarped.
One end should be scribed off with a steel square and cut
off true and square. Use a fine toothed saw to insure
that the cut will be clean and sharp.

Now get out two identical pieces of oak, ¾ in. by
3 in., 30 inches long, and be sure the edges are planed
square and straight. Place them together, side to side,
and square a pencil line across the edges 10 inches from
one end. These pieces are now nailed on either side of
the 2 in. x 6 in. with the pencil lines coinciding with the
squared end, and at the required angle of 5 degrees,
or a 1 in. rise in 10 in. of run. These side pieces must
be exactly parallel and even if you have measured accu-

rately it is wise to sight across them from the side to be sure they are aligned.

The jig should now be mounted on two saw horses so it cannot shift, by toe-nailing the edges of the 2 in. by 6 in. The ends of the two planks that are to be scarfed are cut to the approximate angle or taper with a rip saw, and you don't have to be too accurate because the actual finishing is done with a plane. Lay one of the planks on the jig with the scarf end advanced slightly, be sure it is lined up straight, and set up a C-clamp about 3 feet back of the scarf. Since the plank is probably much longer than the jig, it should be supported at the far end by another horse.

Now the scarf should be planed carefully to a perfect surface and a feather end. A long plane is necessary— either a Fore No. 6, which is 18 inches long, or a Jointer No. 7, which is about 21 inches. The plane iron must be set for a very fine cut, and it should be honed to a very sharp edge. As you can see in the illustration, the plane is held diagonally, with the toe on one side piece and the heel on the other, but the direction of the cut is straight down the scarf. Thus you make a slicing or skewed cut, which is the most efficient way to plane a finished surface with thin shavings. Take heed—reserve your last few cuts for the feather end, which at an angle of 5 degrees is very delicate and easily spoiled. In other words, *finish* from the back forward to the end.

This is a job that requires skilled craftsmanship, but if your plane is razor sharp and you take your time, you are bound to get a good joint. The proper use of the jig will insure both plank ends being identical. There

is an old fashioned way of checking any hand-planed surface for trueness and accuracy. Stand with the work between your eye and the light source, and hold a steel
try-square on the surface. Move it about in various positions, and if no light shows through under the square you
know you have a good job. High spots or hollows become
very obvious with this treatment.

For gluing up a plank scarf you need a proper place
to work—a long, straight, smooth surface that is well supported throughout its length. The simplest arrangement
is a 2 inch plank, 10 or 12 feet long, nailed to a couple of
saw horses. One edge of the plank, the working edge,
must be absolutely straight and clean. Lay the two planks
on this working surface with the scarf in the center.
Support the ends with more horses, boxes or any other
convenient arrangement, and shim them if necessary, so
that the two planks are in a reasonably straight line.

Now make a dry run. Place the *bottom* plank on the
2-inch plank so that it is even with the working edge, slip
a piece of waxed paper under the feather end, and secure
it with a C-clamp near the scarf. Be sure to put a block
under the clamp so the wood will not be marred. Next
lay the top plank in position, flush with the edge of the
2-inch plank. Now here comes the delicate operation.
Because of the long taper of the scarf, and the feathered
end, it requires careful attention to bring the two parts
together exactly right. The best way is to hold a steel
square on the bottom plank while you slide the top plank
along till it meets, as in the illustration. Holding the
plank in position, make a pencil mark across the edges of
the scarf. When you come to glue up, line up these

marks and you'll know the planks will come where you planned.

You will need at least 4 clamps—2 for the scarf itself, and at least one on either side to hold the planks in position. You will also need a piece of scrap, preferably oak, about ¾ in. by 4 in., a little longer than the scarf. This piece goes under the clamps and over the scarf, to distribute the pressure evenly. Another piece of waxed paper goes over the scarf end. Waxed paper is always used when gluing to prevent sticking.

Check the temperature for the ideal 70 degrees, mix enough Cascophen, Weldwood or Cascamite, and coat both surfaces of the scarf evenly. Now place the two planks together, see that the pencil marks coincide, and clamp each plank securely *beyond* the scarf. Sight along the edge of the assembly to be sure the planks are in a straight line. With everything lined up and clamped so they cannot shift, lay the waxed paper over the glue joint, then the ¾ by 4 in. piece so that it evenly straddles the joint, and put on two clamps firmly. A thin bead of glue should slowly squeeze out of the joint, but if it doesn't, it's a sign that you haven't used enough glue! Once glued up, *leave it alone* for at least 10 hours. Don't come around in an hour or so and give the clamps an extra turn or two. You only interfere with the natural curing of the glue and gain exactly nothing.

SCRIBING, SPILING AND TEMPLATE-MAKING

Scribing and spiling are essentially the same thing—the laying out of the line to which a board must be cut to

fit a specific curve. A purist might distinguish the two
in speaking of planking a hull, in that a plank is *scribed*
at the end to fit the curve of the rabbet in the stem, while
it is *spiled* along the edge to fit against the next plank
above or below. Whatever you call it, the same method
is used—the curve that exists is transferred to the board to
be cut with the aid of a pair of dividers or a pencil compass.

   This is one trick that the amateur often fumbles—he
just can't seem to learn to fit a straight piece of lumber
to a member that is curved, and often with a changing
bevel to boot. This one thing, more than anything else,
is responsible for the poor joinery so often associated
with amateur-built boats. Actually, it is relatively simple,
if you have a clear understanding of the method and
work with precision. I am going to try to reduce the
procedure to its simplest form, as clearly as I can, and
believe me—once you have it fixed in your mind, you
should be able to apply it to practically every part of a
boat's construction, whether you are planking, fitting a
cabin trunk, a bulkhead, or a platform for the "john."
In fact, most all of the fitting that goes into the building
of a boat involves the use of this simple skill.

   In the drawing, I have shown two elements—the curved
line at the left, which might represent the rabbet line
of the stem, the crown of a deck, or the inside of the hull
planking; and the square end of a board at the right,
which might also be a bulkhead, a hatch coaming or shelf
in the galley. The series of dotted lines at regular inter-
vals are merely reference points, but note carefully that

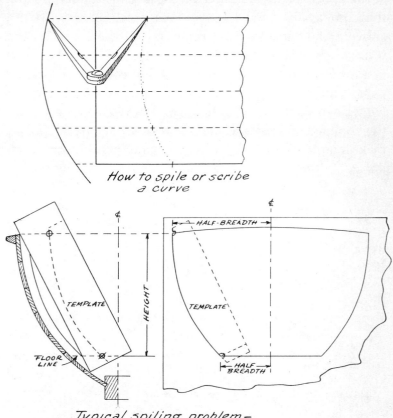

How to spile or scribe
a curve

FLOOR
LINE

TEMPLATE

CℓL

HEIGHT

CℓL

HALF-BREADTH

TEMPLATE

HALF-
BREADTH

Typical spiling problem—
laying out a bulkhead
with a template

they are *parallel* to the top and bottom edges of the board.

With the board end placed close to the curved line, the compass is opened up to span the distance (with a little extra to spare), from the curve to the board on the line which is a continuance of the top edge, and is then clamped securely. Now place one leg on the line of the curve, and tick the board at the top edge with the other leg. Then shift down to the second dotted line, and with the compass legs still parallel to the top edge of the board, make another tick mark. Do the same on the other two dotted lines, and finally a tick on the lower edge of the board.

Now, a line drawn fairly through the tick marks on the board *must* be a duplicate of the curved line on the left. The greater the number of reference points you tick off on the board, the fairer the line and the more evident it becomes that it *is* a duplicate. To carry the thought still further, you can dispense with the tick marks altogether. This time use a *pencil* compass. Start as we did before, with the point of the compass on the curve, and the pencilpoint on the board. Now draw the compass downward, with the point always on the curved line, and holding it always in the same position: so that an imaginary line from the point to the pencil is constantly parallel to the top edge of the board. The result is a continuous pencil line on the board that is exactly the same as the curved line on the left.

This basic description is as elementary as I know how to make it, but it is the sole principle involved in all scribing and spiling, wherever it may be required. You can't undertake boat carpentry successfully unless you know how, so before spoiling some good lumber I strongly urge you to practice. Lay out a curved line, or better yet, spring a light batten on the floor and nail it down. Lay a scrap piece of board on the floor next to it and practice the operation as I have described. If you are at all skeptical, you will at the least prove whether I am right or wrong.

The dotted lines I have shown on the drawing are only for aid in demonstrating. You will sooner or later learn to do without them, but it will help at first for you to draw a series of parallel lines on the board as a guide. A professional merely draws the compass down the curve in one steady sweep, with the two points always parallel to the *imaginary* lines.

To top this little discussion off, let me impress upon you where the novice invariably goes wrong. The most common mistake is to hold the compass constantly *perpendicular to the arc of the curve* (if it is a true arc of a circle), or to every point on the periphery of an irregular curve. This is deceiving when fitting to a very slight curve, for the line obtained may seem right to the eye, but when the board is cut to the line you'll soon discover that it does not and cannot fit.

So now let us get to a specific problem and see how scribing is employed. The next drawing shows a half-section of an open boat—let's assume it is a round bottom dinghy or launch. The problem is to fit a plywood bulkhead athwartships at this point, which is near the bow. It is obvious that you can't get the sheet of plywood into the boat for scribing, so you make a template or pattern of the curve of the side. For templates you use heavy cardboard or wallboard; "Beaverboard" is excellent and very inexpensive.

Cut a straight-sided piece much as I have indicated, and place it next to the frame, with each end resting against the planking, the lower end extending below the floor timber. Now take your pencil-compass and scribe the curved line of the side just as I explained before. Mark the point of intersection with the top of the floor timber, and the sheer, or top edge of the hull at the side. If you are satisfied that you have duplicated the curve of the side, take out the template and cut it on the bandsaw to the *outside* of the pencil line.

The next drawing shows the bulkhead laid out on the plywood, with the center line, height and half-breadths laid out, and the template in position for marking the curve of one side. Flop the template over and mark the other side. Now we'll assume the bulkhead marks the beginning of a small forward deck. If so, the deck line should be laid out, with the proper crown, as I have shown. After re-checking your measurements, the bulkhead can be sawn out. *Don't saw right to the line—allow for fitting.* Keep outside of the line, 1/16 in. or so.

Now try the bulkhead in the boat. Note where it barely fits and where the most should come off. Then trim it down where needed with a block plane, a little at a time, and with several trial fittings. Remember it is easy to cut wood off, but you can't put it back. When fitting *any* member in a boat, small errors in measurement or in scribing are bound to occur occasionally, so always check your work by trial fittings.

It is a good idea to get in the habit of making templates for practically everything. To the enthusiastic and impatient amateur this often seems like a waste of time, for he doesn't realize the errors in marking and scribing that are bound to creep in. So he blithely draws a line across his fine 12 inch plank of mahogany and saws it off. He remembers that it must be beveled, and then it dawns on him that the beveling will leave the plank precisely ¼ in. too short to fit. So the spoiled plank is ruefully laid aside with the fervent hope that it can be used somewhere else, without too much waste.

The place to make your mistakes, and to do your thinking is on the templates, not on your expensive lumber. Much of the fitting in boat carpentry is done by eye, by trial and error, and it is seldom that you can hit it right on the nose in the first try. Even the skilled professional will not attempt an intricate piece of fitting without first making an inexpensive template.

As a further aid in avoiding mistakes, get in the habit of identifying both the template and the lumber. By this I mean a penciled mark designating which side is the face, which end is right or left, and where it is beveled. In cutting and fitting it is often confusing to try and remember just how you planned the piece to go. It is very easy to cut a bevel on the wrong side or end unless you make a penciled note when you first lay out the work.

End Grain

It is one of the cardinal principles of good boat joinery that you must not leave the end grain of a plank or timber exposed. To abide by this rule requires careful planning and some knowledge of the traditional ways in which the professional boat builder solves the problem. The use of light facing strips, moldings, rabbeted joints and corner posts are but a few of the tricks he employs.

The reasons for avoiding the raw ends of timbers or planks are several. In the case of heavy keel and dead-

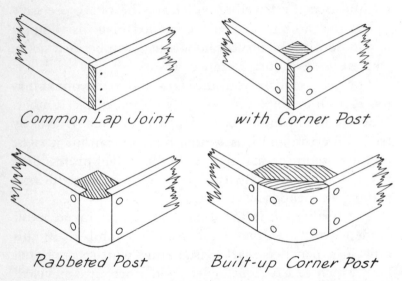

Common Lap Joint   *with Corner Post*

Rabbeted Post   *Built-up Corner Post*

wood members it is to prevent the water from soaking into the end grain. In deck and cabin joinery it is more a matter of appearance. Where the surface is painted the end grain can seldom be hidden. But when you are dealing with brightwork, or varnished mahogany, end grain stands out like a sore thumb and is very unsightly, no matter how you treat it. Oil stains, fillers and varnish bring out end grain about ten shades darker than the surrounding side grain.

In the four accompanying illustrations I have shown typical *corner constructions*. They do not represent any specific application, but it might be a cabin trunk, a cockpit coaming, a berth facing or a counter. The first sketch shows the house carpenter's simple lap joint, which is highly impractical and totally unsuited for boat work. In the first place, the fastenings are driven into end grain, and have very little holding power, whether they be nails or screws. The only fastenings that will hold in end grain are wood dowels set in waterproof glue. Secondly, the end grain is exposed. Further, the sharp corner, since it is end grain, would not last long—with normal wear and tear the wood would soon be splintered off.

In the next sketch we have an improvement—a corner post has been added and we have something solid to fasten to, with wood screws counterbored and bunged. This makes a strong construction, to be sure, but we still have that unwanted end grain.

So in the third sketch we have a *rabbeted* corner post, wherein both problems are licked—it has ample strength in the fastenings and the end grain is entirely hidden. The corner post is first gotten out, rabbeted and fitted *in the square,* and then the outer corner is carefully rounded with a block plane to fair in smoothly with the side planks. The size of the corner post should be worked out seriously on paper first, to be sure to have enough stock to take the proper size screws.

The fourth sketch shows a useful variation. There are occasions where it is desirable to have a much greater radius to the corner. Here the corner post is in two pieces, the outer, filler piece being fitted last, and screwed

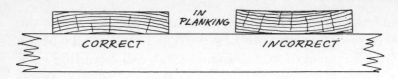

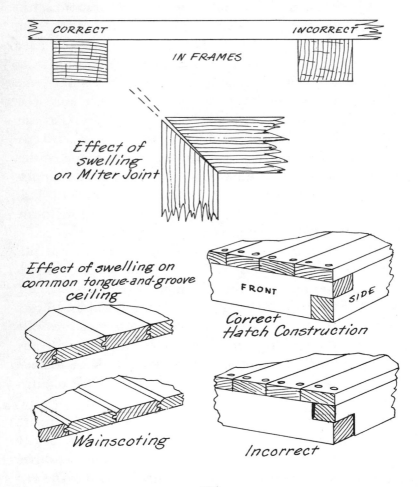

Position of Annual Rings

IN PLANKING

CORRECT   INCORRECT

CORRECT   INCORRECT

IN FRAMES

Effect of swelling on Miter Joint

Effect of swelling on common tongue-and-groove ceiling

FRONT   SIDE

Correct Hatch Construction

Wainscoting

Incorrect

108

and bunged to the inner. This construction avoids the necessity of using very heavy stock, which would be required if it were in one piece. This too should be carefully laid out on paper to insure a well-fitted job. I consider it most important to set corner posts in waterproof glue, to forever prevent water from seeping in and starting the inevitable rot.

WARPING, SHRINKING AND SWELLING

All of the wood used in the construction of a boat will shrink, swell or warp to greater or lesser degrees, as the moisture content varies. It shrinks as it dries, and swells as it absorbs water, and it will warp according to how the grain lies. Since this has a direct bearing on the strength of the fastenings and the tightness of a joint, it is important that you give careful consideration to the grain of the wood before cutting and installing it.

Wood shrinks or swells most in the direction of the annual growth rings (tangentially); only half as much across these rings (radially), and practically not at all along the grain (longitudinally). The illustration shows two planks as they lie, ready to be fastened off. Notice that the left hand plank is placed with the annual rings curved *downward*. This is the correct position for *all* planks, whether you are laying a deck, making cockpit seats or planking a hull. The right hand plank has the annual rings curving upward, which is incorrect, for it will "cup" or warp, and possibly pull the fastenings. If a boat were planked in this manner the bottom could not possibly be smooth—it would resemble a washboard.

The next illustrations show what might be a rib, deck beam, or any framing member whose width and thickness are of approximately these proportions. The left hand frame shows the correct position—with the annual rings more or less parallel to the face. The right hand member has the annual rings standing nearly vertical to the face, and is what is known as "edge-grain." This is bad for two reasons. First, it is very liable to split at the fastenings—not necessarily immediately, but eventually. A nail driven into the edge-grain always tends to take the path of least resistance, which is along the line of the annual rings. Secondly, wood is very difficult to bend edge-wise to the annual rings, no matter how well it is steamed. So remember to look at the end-grain and see that the annual rings are at right angles to the direction of the bend.

The amount of shrinking, swelling and warping to be expected depends on the species of the wood. Generally speaking, the hardwoods head the list. Thus, oak warps, shrinks and swells the most, while the cedars are the least susceptible. It should also be noted that the width of a plank governs the amount of shrinking and swelling—the wider the plank the greater the change. Thus, the amount of swelling across a 12-inch plank is greater than the *combined* amount in three 4-inch planks. This is one reason why hull planking is done with narrow strakes, even in the largest craft.

When a boat is being built under cover, the moisture content of the wood is probably around 15% to 18%, on an average. But when the boat is afloat it will soak up moisture to an amazing degree, possibly to as much as

25%, not only through immersion, but from the higher humidity of the atmosphere. This means the wood is going to do a lot of swelling—and you had better keep that fact in mind as you plan your work. In planking, the swelling of the wood helps you achieve a watertight hull. But in the interior cabinet work the reverse applies, and you have to *allow* for the swelling, otherwise you'll have drawers that stick, doors that won't open, paneling that bulges and removable floorboards that won't remove.

As one example of how swelling can spoil your joiner-work, notice the illustration of a common miter joint, favorite of the house carpenter. As the wood swells the joint opens up in a 'V', and the wider the members the more noticeable it will be. While the miter joint may be used in certain places in the framing of a boat, you will rarely find it in finished cabinet work where it will be exposed to view, for it will always be unsightly and unprofessional.

For another example I have illustrated the right and wrong method of hatch construction, and I think they pretty well explain themselves. The wrong method shows how the swelling of the plank at the edge of the hatch pushes the side frame out of line, cramping the fastenings and opening up the half-lap joint. Such is the power of wood when it swells.

Next I have shown what happens to the tongue-and-groove joint with swelling. Washboard best describes it. Let me hasten to state that tongue-and-groove stock is occasionally used for ceiling in a boat, but in this instance it always has a V-ed joint, (what the house carpenter calls *wainscoting,*) which eliminates the tendency to wash-boarding.

It is neither possible nor necessary for me to illustrate every type of joint used in boat construction. All I have attempted here is to emphasize the importance of giving strict attention to the grain of the wood as you use it, and to the inherent instability of wood in the presence of moisture variations.

USING THE HANDSAW CORRECTLY

The most noticeable difference between the work of the average amateur and the skilled professional is in the quality of their joints, the precision with which two pieces are cut and fitted together. When the surface is to be painted, rough edges and voids in a badly fitted joint can be concealed with trowel cement. In fact, amateurs often depend on it to hide their poor workmanship and mistakes, instead of improving their skill to the point where concealment is unnecessary.

But when it comes to bright-work, where the clear transparency of varnish reveals every scratch, ragged edge and open joint, poor workmanship stands out nakedly for all to see—a constant reminder of the inept worker. Trowel cements and surfacing putty may be had in a mahogany color, but they will still be seen through the varnish, and only serve to make the poor work a little less noticeable. So a conscientious worker, who takes pride in accomplishment, applies himself diligently to the task of learning to work accurately, cleanly and with precision, so that he has nothing to hide.

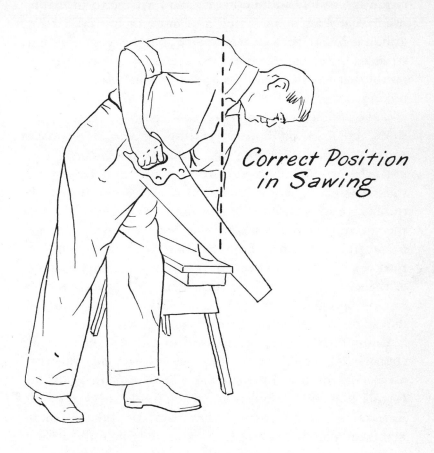

Correct Position in Sawing

Clean, accurate cutting with a handsaw requires lots of practice, particularly in fine cabinetwork where precision is absolutely imperative. Now you have probably used a handsaw many times, and more or less take it for granted, but have you ever really checked your method of using it, to see whether you are using it properly? I doubt that you have, and I suggest you make a simple test to find out.

Take a piece of scrap—any odd board at least ¾ in. thick and 6 in. or more in width. Draw a line square across it near the end and put it on the saw horse as you normally would in position for cutting, and have your steel try-square handy. Now start sawing the board off to the line in your *normal* manner of working. Stop the saw in mid-stroke, holding the pose without moving. Look straight down and notice where your right shoulder is (if you are right-handed,) in relation to the cutting edge of the saw. If it is *behind* the cutting edge, you are the typical amateur, sawing incorrectly! The shoulder should be *above* of the cutting edge.

Resume sawing and again stop in mid-stroke, without changing the position of the saw. Now place the try-square on the board next to the blade, and the odds are 10 to 1 that the blade is not at right angles to the board, and you are *undercutting*. This is a natural tendency, and takes a lot of practice to overcome. The only answer is to check with the try-square for plumbness every time you use the saw.

Notice the stroke you use when sawing. Most amateurs use short, jabbing cuts, thus doing all the cutting with the middle section of the saw blade. I shall always remember my father's admonishing, "Won't you ever learn that a saw has teeth on the ends as well as in the middle?" The proper stroke is a long, steady one, utilizing the full length of the saw. This gives you better control—a straight cut and less wobbling.

For clean cuts where a fine finish is required, a fine saw is necessary—not less than 11 teeth to the inch—and it *must* be sharp, with just enough set to prevent binding. Now I would like to point out that even such a saw does *not* make a really sharp-edged cut. The teeth actually tear the wood fibers apart, as you well know when you look at the underside of a cut. They tear the wood on the upper side or face of the board also, but to a lesser degree. This is most noticeable when sawing soft woods.

But when you are working on interior fittings or trim that is to be varnished, your joints should be as inconspicuous as possible, which means no ragged edges on your saw cuts. Here is an old cabinet maker's trick that makes the job easier. Instead of laying out your cuts with a pencil line, use a steel straight-edge and score a line with a sharp penknife, as shown in the illustration. Bear down firmly and make a clean scribe—it need be no deeper than 1/32 in.

Now here's the tricky part. You must saw very accurately *outside* the scored line, but close to it, so that the inner side of the saw teeth just barely miss it. Try it with a piece of scrap wood first, and notice that the normal tearing of the wood on the face *cannot* go beyond the knife-cut.

Thus you will see that the finished edge is sharp, clean and precise. Obviously, you'll need to handle the piece

Scoring wood for clean cut

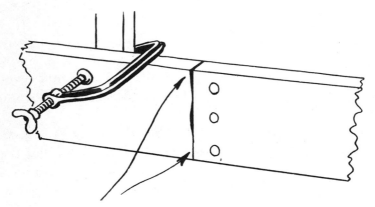

Run Saw through joint
for tight fit

with care when handling or installing it, for the sharp edge can be spoiled very easily.

Here's another trick that is very commonly used when fitting two plank-ends in a common butt-joint. Let's assume that the plank on the right, "A," has already been fastened off. When you bring the other plank, "B," up to "A," you discover that the joint fits badly and stands open as shown.

Clamp the plank "B" in position, as close as it will bear. Now run your cross-cut saw down through the joint. Remove the clamps, bring the plank up again, and you'll have a well-fitted joint, wood to wood.

VI

About Fiberglass

In the building and maintenance of wood boats there are instances where the judicious use of glass-reinforced laminates can be advantageous. Fiberglassing adds considerably to the strength of wood, and can greatly reduce the normal wear and tear from abrasion and exposure to the elements. But before discussing the various applications, it is advisable to determine the characteristics of the materials, and particularly their limitations. Because of their relatively high cost, failures resulting from improper use can be very expensive.

Fiberglass cloth has great strength and abrasion resistance, and is very stable. In fact, it has practically no elasticity, and this is a point to remember, as we will elaborate on later. It is also quite heavy, and adds considerably to the weight of a small boat such as a dinghy.

Polyester resin is strong, stable and adequate for most applications of fiberglass, or where the demands are noncritical. But where great strength and maximum adhesion are required, as in bonding to very hard, dense woods and metals, epoxy resin is far superior.

Epoxy resin has a very high tensile strength—much

greater than polyester, and is the strongest adhesive that can be made. It is handled in the same manner as polyester, but requires higher temperature and a much longer curing time. Since the best working temperature for epoxy is 80 to 85 degrees, it is obvious that some form of artificial heat, such as heat lamp, is necessary. It can be used as a base coat on difficult surfaces before applying polyester, but you cannot use epoxy over polyester.

As an alternate reinforcement, Dynel may be used instead of glass cloth. A product of Union Carbide, Dynel is a fabric of synthetic, vinyl-acrylic fibers. It is much lighter in weight than fiberglass, has a soft, fuzzy character, and can be stretched around corners or compound curves without wrinkles. It "wets out" quickly, and absorbs a great deal more resin than fiberglass.

It is very easy to use, for it is applied dry—that is, it is stretched over the surface tightly and stapled in place, then the resin is applied to the dry cloth with a roller, brush or squeegee. It readily soaks right through to the wood, and fills the cloth completely. Two or more coats are required, depending on the finish required.

Where to use fiberglass

Many an amateur, having an old, carvel-planked boat with gaping seams, has turned to fiberglassing the hull as a means of making it watertight, and invariably it has resulted in complete failure. My advice is DON'T!

When the planks swell and shrink, tremendous forces are exerted in several directions, and something just has to give. Since the resin-impregnated cloth has absolutely no elasticity, it will crack and split along the seams, and undoubtedly delaminate. The same thing will happen with strip-built decks. The whole trouble lies in trying to bond a rigid material to an unstable surface that is constantly working and flexing.

The ideal place to use fiberglass is on plywood, which is a stable material not subject to expansion and contraction. All joints should be taped, of course, to guard against possible working of the fastenings. In preparing the surface for fiberglassing it is necessary to fill the joints, dents or gouges, and over the heads of fastenings. NEVER use oil-based compounds—use only water-based or the epoxy-type compounds.

One rule to remember is never apply fiberglass to a

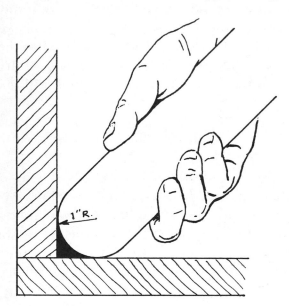

FILLETING A CORNER

sharp, "inside" corner. As an example, if you are fiber-glassing a deck the cloth should be turned up along the cabin trunk and coaming. If you were to turn the angle there, there is a strong possibility that air will be trapped under the cloth in the corner. This results in poor adhesion, possible delamination, and loss of strength.

The proper method is to form a curved surface in the corner by means of a fillet. While the fillet may be of wood or water putty, a far better way is to make a thick mixture of chopped glass fibers and resin. With a template of sheet metal cut to the required radius, the mixture can be troweled in neatly. Thus the fabric will turn the corner in a smooth curve.

If the surface has previously been painted, it should be sanded off to bare wood with a very coarse paper, and the disk sander is best for the job. NEVER remove the paint with a blow torch or paint remover—they only drive the oils and paint into the wood, and the resin will not bond properly. After sanding, with the wood considerably roughed up, it is a good idea to wash it down with acetone or a strong detergent—and allow plenty of time for drying out.

In fiberglassing small craft, such as a dinghy or one-design racing sailboat, I would prefer Dynel over glass cloth. In the first place, it means a considerable saving in weight, costs less, and has a higher resistance to abrasion. Small craft are continually being dragged on the beach, over sand, gravel and rocks, and Dynel gives far greater protection than glass cloth.

One useful application of fiberglass is in making a water tank. It can be made whatever shape is necessary to fit

the space allotted, at a fraction of the cost of a custom
made stainless steel or monel tank.

The illustration shows construction details of a simple
tank such as might be fitted under a side deck. It is not
to scale, and no dimensions are given, since it is merely a
suggested method.

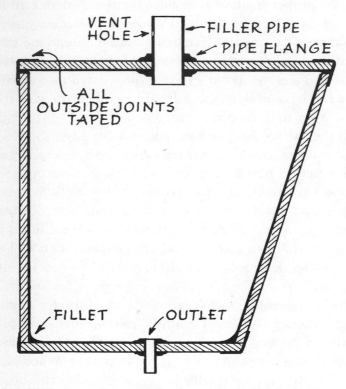

Having determined the exact size and shape of the tank,
you construct an open box, using ¼ in. plywood, with
glued joints. All inside corners must have a fillet.

The inside of the box must now be fiberglassed. Polester resin can be used, and you'll need enough for four coats—base coat, saturation coat, and two finishing coats. Any seams in the cloth should have a 2 in. overlap. Note that the cloth must be carried *over* the top edges of the box.

The top of the tank is a piece of plywood cut to fit accurately, fiberglassed on the under side. When it has cured, the filler-pipe is installed, using two pipe flanges screwed on tightly. I would use some thiokol-type compound under the flanges. The outlet pipe is installed in the bottom in the same manner. You are now ready to put the top on.

The top is put on with resin as the adhesive—coat the top edges of the box and the under side of the top liberally. Then cover *all* joints with 2-in. glass tape and four coats of resin.

If the tank is to be used for drinking water, you may find it takes some time to get rid of an odd taste. I would let the tank cure for at least a week, then flush it out with a strong detergent and hot water. A hot solution of baking soda is an alternative.

To the professional, this tank would look pretty crude, and so it is. But it is inexpensive and relatively easy to make. Actually, a fiberglass tank, properly made, doesn't require a plywood covering. Three layers of glass cloth, laminated, would give a wall thickness with sufficient rigidity to eliminate the need for outside support.

Fuel tanks can also be laminated fiberglass. But take warning—the design must meet the requirements of the insurance underwriters and the Coast Guard!

In this method, the plywood box serves as a *female*

mould for the lamination. After the corners have been filleted, the inside of the box is given a heavy coat of melted paraffin, applied with a brush. This acts as a parting agent, so the laminate can be removed from the box.

I would recommend using 12½ oz. fiberglass, and three layers. Epoxy resin should be used for the added strength that is required. The top laminate can be made easily on a piece of plywood coated with wax.

When the fiberglass has properly cured, you may have difficulty removing it from the box, unless all four sides taper toward the bottom. Care and patience are in order.

As in the other type, the top is secured in the same manner with the seams well-taped.

VII

On the Use of Wood Preservatives

One of the hardest things to explain is why a seemingly well-built boat just a couple of years old can develop a bad case of dry rot, while many a nondescript old tub will be sound as a dollar forty years after her initial launching. I am sure I don't know the answer, nor have I found anyone who does. I suppose boats are like people —some get poison ivy and others are immune. But I do believe we will see the day when dry rot is far less common than it is at present, through the further development of effective wood preservatives.

We know that dry rot is caused by a fungus—and so is "athlete's foot." A fungus is developed from spores, which thrive on dampness, or a combination of heat, humidity and lack of ventilation. Now these spores aren't just "born" where they appear—they are carried from somewhere else in the air or by contact, and reproduce and grow where conditions are favorable. We know that athlete's foot is contagious, that it is transmitted from one person to another. So it is logical to assume that dry rot is also contagious.

Lest this analogy seem far fetched, let me say that I

know of more than one boatyard in which rot-infected wood that has been removed from a boat is immediately burned, to prevent the spores from spreading to the nearest pile of lumber, or to adjacent boats. Indeed, some people take this matter of rot very seriously, as well they should.

The insidious part of dry rot is that, like cancer, it is seldom discovered until it has gained a foothold and damage has been done. It has a habit of starting in the most inaccessible places—probably because they are the least ventilated—where it cannot be got at without literally dismantling the boat. It is a costly, hideous disease, but like cancer, it looks as though it may some day be preventable.

In the past fifteen years or so, considerable progress has been made in the development of wood preservatives. Today, the treatment of wood in boat construction is common practice in all of our good boat shops. Belief in its effectiveness is widespread, and there is as yet no evidence to the contrary.

However, no one yet knows just *how* effective wood preservatives are. No one can say that one treatment will forever prevent dry rot from attacking a boat. Twenty-five years from now we may know, but in the meantime all we can do is use the preservatives liberally, and hope for the best. It is known that preservatives do *arrest* dry rot, and prevent it from spreading. It is also known that they *will* prevent it from starting—but for how long, no one has yet determined. Some say a year, at most, that the chemicals will gradually leach away and leave the wood unprotected. Others will say no, once the chemicals are in the wood cells they are there to stay, unchanged and effective.

I wouldn't know, and I doubt anyone does as yet. Only time will tell. In the meantime, it is my belief that the best policy is to literally drench every piece of wood that goes into a boat with wood preservative. Certainly it can do no harm, and if it delays any nearby fungus spores from setting up housekeeping in my boat for so much as one year, I'll consider it worth the effort.

There are two types of preservatives commonly in use at present. One is *copper napthanate,* the other is *pentachlorophenol.* In both types the basic ingredient is in suspension in a volatile petroleum solvent, which gives the mixture its ability to penetrate.

The copper napthanate type of preservative is made in two forms—one is a clear green in color, and the other is colorless. The green kind is used principally for treating the frames and timbers of a boat. It stains the wood green, leaves the surface quite sticky, and will bleed through the prime coat of paint. The clear kind can be used under varnish or paint without affecting the color of the wood.

The pentachlorophenol type often contains other chemicals, such as phenyl oleate, tetrachlorophenol, and a couple of other jawbreakers that are even longer. I know practically nothing about chemistry, and can't tell you what these ingredients are or what they are supposed to do. All makes of preservatives have their components listed somewhere on the can, in case you are interested. All makes of this type of preservative that I have used are colorless.

While I don't want to step on the toes of any manufacturer, I have used both types of preservatives many times, on all kinds of wood, and I honestly must say that I believe the penta etc. type penetrates the wood deeper and more quickly than the copper type. Let me hasten to add that it may not be any more effective than the other in preventing rot, and its superior penetration may be entirely due to the petroleum solvent used as a vehicle. I just happen to like it better.

Regardless of which brand of likker you prefer, the treatment is the same. You apply it liberally to all surfaces of the bare wood with a brush. Naturally, it cannot penetrate a painted surface. The wood must be dry, for it can not penetrate if the wood is already full of water. Remember too, that the wood cells run in a longitudinal direction, with the grain, and therefore the greatest absorption is in the end grain. So for maximum penetration the wood should be stood on end in a pan of preservative. Since this is not always practical, be sure the end grain gets a good dose with the brush.

Since the clear preservatives are colorless, and the petroleum solvent vehicle is so volatile, you cannot tell, a half-hour later, where you started and where you left off. There is no visible evidence that the wood has been treated. So if you are treating an entire boat and have to quit before it is finished, make a careful note of where you stopped.

Not all makes act the same way. Some appear to be dry in less than an hour, others leave the surface with a slight stickiness for 24 hours or more. Needless to say, some trace of the chemicals remain on the surface of the wood, even though you may not see or feel it. Therefore when gluing wood that has been treated, regardless of what kind of glue you are using, you *must* wait at least 24 hours, and you *must sand the surfaces* well, or the glue will not hold.

Likewise in painting or varnishing over treated wood, wait at least 24 hours, or you may find it won't dry, or if it does, it will peel sooner or later.

There is another aspect to the use of preservatives that I neglected to mention. We know that the fungus of dry rot cannot thrive in the presence of the active chemicals which are the base of these preservatives. And we also know that the fungus thrives in dampness. Now, wood that has been adequately treated with preservative, and by adequate I mean well soaked, actually absorbs much less water than untreated wood. So the treatment is a double-barreled one, and must be very frustrating to the fungus.

So far we have been discussing the treatment of new construction, while the boat is a-building. Here it is merely a matter of preventative treatment, and it is extremely simple. But when dry rot is discovered in a boat that has been completed and is in use, things get complicated. First you must, if possible, find out how and why it started. Secondly, determine how you are going to get the diseased member removed from the boat, with the knowledge that once it has been removed you may find the rot has spread to adjoining members. Thirdly, treat the entire area with preservative to kill any existing spores, and prevent the rot from starting anew. Fourthly, replace those members that have been removed with pre-treated, sound, new wood.

Taking these problems in their order, it requires quite a bit of detective work to determine the cause of a specific case of dry rot. No two boats are constructed alike, and rot may start anywhere. Nor is it always immediately visible. Where you only *suspect* its presence, you can often learn a lot by sounding with a hammer. Sound wood has a sharp, hard ring when struck or tapped with a hammer. Diseased wood responds with a dull, mushy thump, gives no "bounce" to the hammer, and it feels as though you were striking a block of cork. Dry rot turns the wood to a soft, crumbly punk that can be dug out easily with the fingernail, and you'll have no difficulty recognizing it at first sight. It makes me shudder just thinking about it.

Much dry rot is caused by rain water seeping into a hairline crack, working down between two members where it is trapped so that it cannot evaporate. This spot may often be a considerable distance from the point of entry. If and when you locate the crack that is responsible, the first thing to do is to pour lots of preservative into it, and then seal it with a good seam compound or trowel cement.

As to the removal of the infected wood, specific instructions obviously cannot be given. You should cut out not only the affected part, but a considerable distance beyond its limits. Be suspicious of all members in the immediate vicinity, and remove as many as are feasible for inspection. Remember, rot is invariably more extensive than is first suspected.

To cite a specific case, I found a small spot of dry rot the size of a silver dollar in my boat, in the coaming or trim of the cockpit. I removed the whole 5-foot piece for replacement, pleased that it was so simple, only to discover more rot in the deck carlin. Before I was finished the whole cockpit had been ripped apart, and what appeared to be an hour's job developed into a 2-week project that left me a nervous wreck.

So where it is possible, it is better to replace the whole member than to try to cut out a section and fit in a graving-piece, or "dutchman" as it is called. It's a tough problem, no matter how you approach it. One of the troubles involved is the removal of fastenings, for often they are inaccessible, or on the wrong side of the member. Then you can only split the wood off in sections. Where drifts are encountered, the wood must be split off and a hacksaw resorted to. Considerable ingenuity is called for, and you have to compromise with expediency. In the final analysis, much of this tearing out and replacement is beyond the skill of the average amateur, and you would be well advised to call in a professional who has the experience and know-how in such work.

Once the affected parts have been removed, they should be burned. All surrounding areas should be drenched with preservative for as far as you can reach. All new replacements should be well soaked, *after* they have been fitted, and *before* they are installed.

Once you have been through this truly heartbreaking experience, as I have several times, you will forever after be suspicious, alert, and on your guard. You will be continually sniffing, snooping and probing about the boat, minutely inspecting every inch of the deck, cockpit, cabin structures, stem-head and transom for tiny cracks where rain water or melted snow might enter. Henceforth, when engaged in new construction, or making alterations, you will make danged sure that your joints are really tight, and are put together with waterproof glue or bedding compound, and through it all, your trusty gallon can of wood preservative will be within arm's reach.

There are several products on the market for the repair or treatment of dry rot. In general, a liquid resin is injected into the rotted area which quickly cures and hardens to a solid mass far stronger than wood. It is most effective where the rot is confined in a small section of a plank or frame and the extent or limit of the rot is clearly defined. But where rot has spread from one member to other adjoining members, the only recourse is to surgery.

In my opinion, the best of these products is one called "Git-Rot," made by Aladdin Products, Huntington, N.Y. It is an epoxy resin of low viscosity with extraordinary wetting ability. It soaks in and completely saturates the rotted wood, and hardens to a tough, resilient mass.

It comes in two bottles—the resin and the hardener. Once the two are mixed, the reaction starts, and in $\frac{1}{2}$ to 1 hour it begins to harden. The higher the temperature the faster the cure. In addition to repairing rot, it is an excellent preventative. Any hairline crack, check, joint or seam can trap fresh water and start rot. Because of its low viscosity, "Git-Rot" will penetrate and completely fill even the tiniest of openings, and once it has cured, it is permanent.

MILDEW

The same fungus that causes dry rot also takes the form of mildew, and it frequently attacks the painted surfaces in the cabin. It results from a combination of heat, humidity and lack of air circulation. Some seasons it is more prevalent than others, and most boat owners are troubled by it sooner or later.

The problem is what to do about it when repainting. The unsightly gray mold or fungus that lies on the surface of the paint can be easily wiped off with a damp cloth, but then you'll find the gray spots are actually in the paint itself, and still visible. So you might think that the obvious thing to do is to sand them out. Well, you probably can, but you haven't gotten rid of the mildew, for it will grow and re-appear right through the fresh coat of paint.

It is all right to wash and sand the spots off, but in addition you should *kill* the fungus, for it still exists in the paint, although you can't see it. Now the very wood preservatives that I discussed elsewhere will definitely kill the spores of mildew. So before repainting, saturate a cloth with one of these preservatives and carefully wash down all painted surfaces, whether you can see mildew spots or not.

I don't know whether this is an original discovery or not, but as an experiment, I painted my cabin interior with an acrylic latex house paint, and over a 5-year period I have not seen a trace of mildew. Upon my recommendation several friends have switched to this paint, with the same success. It is my earnest belief that mildew spores thrive in the alkyd-resin-based paints, but not in acrylic.

VIII

Repairs

In the normal use of a boat there is a certain amount of unavoidable wear and tear that can be kept under control by the judicious use of trowel cement, paint and varnish, but not forever. Eventually you are faced with repair work—old wood and tired fastenings must be replaced with new, and the boat restored to its original youthful beauty. In addition to this normal wear and tear, there comes the infrequent unexpected—the damage from grounding, collision, chafe, hurricanes and dry rot.

To give detailed directions for making all the possible repairs that these things might necessitate would be a herculean task, and would fill several volumes. I can therefore only cite a few of the more common instances, explain the procedures, and discuss some of the problems that are common to all repair work.

Repair work is a serious thing, and not to be entered into lightly. What may appear at first glance to be a simple, 3-hour job, can turn out to be a 30-hour major project after it is started. Often you'll strike a snag that you hadn't thought of. Fastenings refuse to come out and break in the wood, the piece of trim you intended to put back later must be split apart to remove it, or the member you thought would come free by removing three screws happens to be edge-fastened to several other members. These are typical of the troubles many amatuers run into by not giving proper thought to the job before it is started.

When a damaged member must be replaced, the first thing to do is to analyze every step of the operation, find out what is involved, and how you are going to do it. Use your imagination, and anticipate every possible trouble. Determine how and where the piece is fastened, and if the fastenings are accessible. Suppose they break or are frozen in the wood. How are you going to treat them? Are you going to have to remove *other* members, because they may be fastened to the first piece, or to get room to work? Decide how you are going to cut the damaged piece out, where you are going to cut it, and finally, how you are going to fit and fasten the new piece.

These things may seem very elementary—indeed, they are meant to be. But I had to make a good many mistakes and spoil an awful lot of nice wood before I learned that the time to make mistakes and to do your best thinking is before you start to work. Meticulous planning beforehand can make any job simpler.

Removing old fastenings can often give you a lot of trouble. Common brass screws de-zincify in the presence of salt water, and generally crumble or twist off. Bronze screws often freeze in the wood, particularly in oak. Hold a heavy screwdriver in the slot and strike the handle with a hammer a few times, and you'll find this may free it. Always remove the screw with a brace and a screw-driver bit—you'll be less liable to damage the slot.

Galvanized boat nails freeze in oak, through oxidizing of the zinc coating, and from rust, and are invariably impossible to pull out. Then your only recourse is to cut them off or drive them the rest of the way in. Gal-vanized iron drifts are even worse, and can't be removed short of splitting the wood off, which of course destroys the timber.

Very often it is necessary to remove a section of trim, or other member, in order to get at something else, and you want to put it back when finished, undamaged. This is a ticklish bit of work that requires extreme care and patience, particularly if the trim member has a varnished finish. You must remove the bungs to get at the screws without marring the surrounding wood, and above all, you must preserve the sharp edges of the counterbore, otherwise you'll have an unsigthly finish when the new bungs are in place.

Here's the easiest and quickest way to remove bungs cleanly. Cut the head off a *steel* wood screw, #12 or #14, and put the screw in the chuck of your hand brace. Cen-ter it carefully in the bung and screw it in SLOWLY. When the point of the screw fetches up on the head of the fastening, the bung will be eased right out!

Most amateurs peck away at the bungs with a small chisel, then switch to a penknife or icepick, chipping out tiny bits of wood like a termite. This is tedious and time consuming, and not always successful. Inevitably the tool will slip and mar the edge of the hole, or old brittle glue is encountered which prevents the complete removal of the bung.

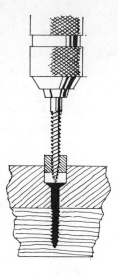

A number of years ago, while the motor was being hoisted out of a sizeable auxiliary, the heel fitting of the derrick-boom gave way as it was being swung shoreward, and the motor struck the cockpit coaming heavily, with the result much as I have illustrated. The cap was splintered and a section broken out, and a severe gash was torn in the coaming.

Now the reason I have shown this is that it illustrates a typical repair job, and the manner in which it was approached. Furthermore it demonstrates a simple method that has many adaptations. At first glance it might be assumed, as the owner did, that the entire coaming would have to be removed and a new one fitted. Obviously the cap *could* be patched, but it was quicker and easier to fit a new one. Once it was removed it was discovered that the coaming was drift-bolted to the deck carlins, the heavy fore-and-aft pieces framing the cockpit.

To have removed the coaming, with the possibility of

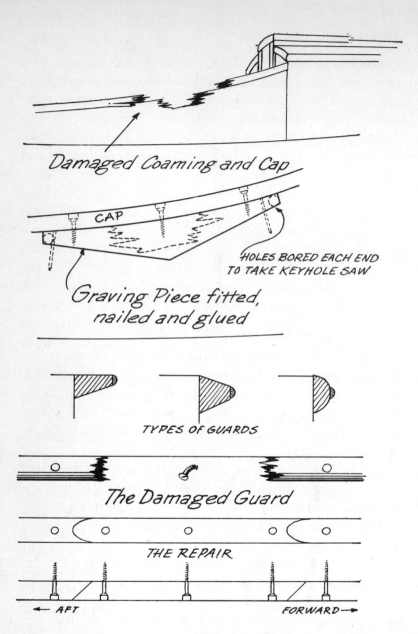

Damaged Coaming and Cap

CAP

HOLES BORED EACH END
TO TAKE KEYHOLE SAW

*Graving Piece fitted,
nailed and glued*

TYPES OF GUARDS

The Damaged Guard

THE REPAIR

← AFT

FORWARD →

not being able to free the drifts, and to make and fit a new one, would have been an expensive and unnecessary project. The solution was to fit a graving piece, commonly called a "dutchman," where the coaming was damaged. Fortunately, there was no drift bolt where the motor struck, or the coaming would have been torn loose or seriously wracked.

The graving piece was fitted as I have shown. First the outline of the shape enclosing the damage was accurately drawn in the coaming. Then two 1 in. holes were bored through as shown, and the piece was sawn out with a keyhole saw. A wide chisel was used to clean out the corners and true up the saw cuts. The mahogany stock used for the repair was selected to match as nearly as possible the wood in the coaming. The shape was taken off by clamping the piece to the coaming and running a sharp pencil around the inside of the opening. After it was carefully fitted, it was fastened with two long galvanized finishing nails as shown, and waterproof glue on all edges. When the new cap was put on, the screw fastenings were laid out so that one came in the center of the graving piece.

Now I want to point out one thing, the shape of the graving piece. Notice its resemblance to a common scarf. The reason it was designed that way was to reduce the length of the cut at the ends, and make it easier to keep the face in line with the coaming. An end-joint crossing the grain of the wood at right angles is always more conspicuous then when it is at an angle. Had the piece been laid out as a simple rectangle, the ends would have been more than twice as deep, there would have been no way to properly fasten the lower corners, and the graving piece would have been much more noticeable when finished.

This, with variations, is one of the commonest methods of repairing a damaged member where the injury is confined to one spot, and where it is not practical to remove the entire piece. I would point out that the "dutchman" must be fitted perfectly, wood to wood, so that the joint is merely a hairline, or it will always be conspicuous—continuously calling attention to the fact that a repair job has been done.

One very common type of damage is an injured guard- or rub-rail, as shown in the next illustration. This generally occurs amidships, or thereabouts, and is caused by laying against a piling, bulkhead or dock in rough water, without proper fenders. Contact with another boat can likewise splinter or break out a section of guard. To give some protection from this, the better built boats will have their guards capped with a brass half-round, or half-oval. Guards may be of various shapes, as I have shown. For the purpose of demonstration I have taken the half-round type.

To cut out the damaged section and fit in a graving piece, first locate the nearest fastenings, which are generally screws, with bungs. The cuts should be made within 3 or 4 inches of those nearest the limits of the injury. Thus, if the guard has been split, the cuts must be made still farther back, where the wood is unharmed. The cuts should be made with a fine-toothed saw, at a 45 degree angle, and the most important thing to remember is that both cuts should run *forward,* from the outer face, as shown. With the fastenings out and the cuts made, the damaged piece may often serve as a pattern for shaping the new piece, in cross-section.

Don't try to dress the new piece to the exact cross-section—keep it a little full, so that you have something to dress down when fitted. Cut one end to the 45-degree angle, then place it under the old guard to check the accuracy. Then take the measurement for length on the *outer* face, and cut the other end, a little full. For a final fit, if the new piece is a trifle long, as it should be, hold it in place and run a saw through the joint, as I described in the section on "Using the Handsaw Correctly."

After the piece has been fitted, the back should be given a coat of paint, and allowed to dry. *Don't* get any paint on the end-grain, for the joints should be made with waterproof glue.

Fastenings for the piece should be laid out with care. Note that the two end ones should not go *through* the scarf joint, but close to it. The natural tendency would be to have the fastening go through both pieces, to tie them together, but actually this would eventually cause the ends to split, and would be much weaker. With the ends glued, the screws in and bunged, the piece can be finally dressed down to the exact shape of the old guard with a block plane, and then sanded smooth.

Another common type of damage is the punctured plank—the hole in the hull from collision, striking a submerged object, or grounding on a rocky bottom. Sometimes this injury may be severe, with several planks and perhaps a rib or two broken. But for an example, I have

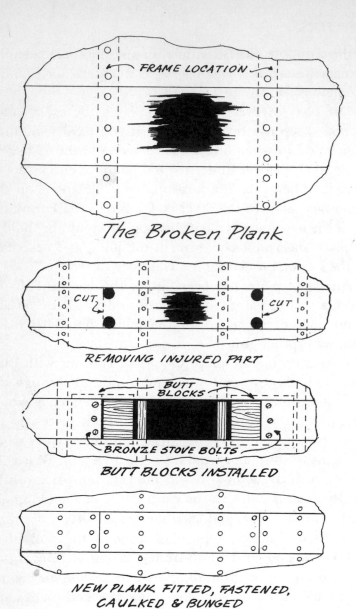

FRAME LOCATION

The Broken Plank

CUT   CUT

REMOVING INJURED PART

BUTT
BLOCKS

BRONZE STOVE BOLTS

BUTT BLOCKS INSTALLED

NEW PLANK FITTED, FASTENED,
CAULKED & BUNGED

shown a simple punctured plank, with the hole between two frames.

Now before you start a job like this, give a thought to several of the cardinal principles of good boat construction. First, the butt joints in hull planking must never land on a frame, but in the center of the frame bay, midway between two frames. The plank ends are screwed to a butt-block, which must overlap the next plank above and below. Secondly, all butts are shifted or staggered, for strength. This means that where a butt occurs in a plank, neither the plank above or the plank below shall have a butt-joint nearer than three frame-spaces away.

In the example, I have shown the damaged plank cut in the next frame bay from the injury. But this might be wrong—if the next plank above or below has a butt-joint in the same place, you must go over at least to the next frame bay to make your cut, even further if practical.

Once you have decided where you are going to cut the plank, draw a line squarely across it, and bore a 1 in. hole through, close to the line and next to the top edge. From the hole cut across the plank with a keyhole saw. When both ends have been cut, take out the fastenings and remove the damaged section. If a fastening refuses to come out, it must be cut off flush with the frame, and you may have to split the wood to do this. Make a note where the cut fastening comes, so you won't hit it when you bore for a new fastening.

With a piece of material of the same wood, the next step is to scribe it for fitting. Often you can do this by having a helper hold it over the opening, while you mark it from the inside of the boat. Or if it is not accessible, take careful measurements with a rule. Cut full of the line and fit it by trial and error. Remember the edges must have a very slight bevel to allow for caulking, and it must be a tight fit, wood-to-wood on the inside, with no daylight showing through.

Now before the new member is installed you must make two butt-blocks. They should be of oak, at least as thick as the planking, the length should be about 12 times the thickness of the planking, and 1 in. wider, so they will overlap the next planks above and below by ½ in. The blocks should be heavily coated with bedding compound when you put them in.

The butt blocks must be passed inside through the opening, clamped in position, and the ends of the existing planking fastened to them with wood screws, counterbored and bunged. The new plank is brought up and fastened to the butt blocks, and to the frames, in like manner.

Sometimes the repair happens to be where there is considerable bend in the planking, and you may have difficulty bringing the ends of the new piece into the butt blocks. If so, don't depend upon the wood screws to pull them in. By the use of shores or braces you can generally spring the ends in fairly close. Use flat-head stove bolts, instead of screws, and the plank end will be drawn up tightly to the butt blocks. The stove bolts should be counterbored and bunged, just as you would with screws.

With the new member in place, all that remains is to paint the seams and caulk them, fill with seam compound, and finally smooth up everything and paint.

A puncture in plywood can be repaired in much the

same manner. A square section is cut out and a new piece fitted, with a backing piece of the same plywood—in effect, a butt block. However, in addition to wood screws spaced fairly closely, waterproof glue should be used in fastening off. The butt block, the back of the new piece, and all edges should be liberally coated, and when completed, any voids in the joint should be filled with glue.

Another type of repair that is not uncommon is the broken frame, particularly in boats that are getting along in years. Frames generally break at the turn of the bilge, where the bend is sharpest, and the wood fibers are severely distorted and compressed. As shown in the illustration, the planks are usually sprung off at the point of fracture, so the problem is two-fold—strengthening the frame structure, and bringing the sprung planks back into line.

The first step is to secure some bending stock—clear, absolutely straight-grained oak that is not too dried-out. The dimensions should approximate those of the broken frame—the thickness should be the same, but the width can be a little greater. It should be long enough to lap over *at least* two planks each side of the break.

Making careful note of the direction of the annual rings in the end grain, the new frame is slit down the center to a point *not less than* 4″ from the end. This is done with a ripsaw, and must be a straight, accurate cut. The main reason for slitting the frame is to eliminate the severe deformation of the wood fibers that probably caused the old frame to break. Thus when the slit frame is bent, the inner half can slide by the outer half with no distortion.

The slit frame should be steamed or boiled before bending, and while it is cooking you can prepare your fastenings. Bronze stove bolts should be used—screws do not have the holding power for this job. The bolts should be counterbored and bunged, as with screws. If

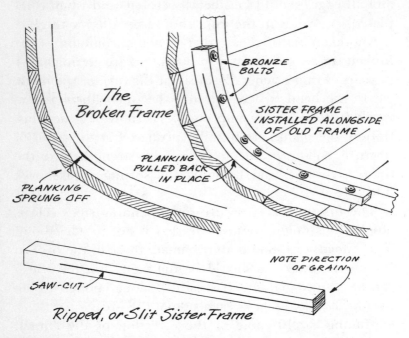

Ripped, or Slit Sister Frame

the planks are sprung off, you should force them back as much as possible by means of shores and braces, rather than relying on the bolts alone to bring them back in line.

As shown in the illustration, the new frame is installed with the slit end down, and the solid end up. This is to prevent water being trapped in the saw-cut. The frame can generally be forced into place with your foot, but it is better to use shores or braces, to hold it in place while boring for the fastening. In boring and setting up the stove bolts, work from the *top downward,* one at a time.

It is unnecessary to fasten the new frame to the broken one—it would accomplish nothing. Just make sure it lies snug against it. When it has finally dried out to normal moisture content, give it a good soaking with wood preservative, and be sure the old frame gets the same treatment, for the fracture is an ideal spot for rot to start. I might add that you'll generally find the plank seams need cleaning out and re-caulking in the area of the damage.

Many an amateur is faced with the problem of restoring a second-hand boat to serviceable condition. Let's assume she is old, but sound. The planking has dried out and shrunk badly, and the old, crumbly seam compound and caulking has dropped out of the gaping seams in many places. How can it be made tight? Driving new caulking in won't work, for the seams are too wide to hold it.

The best answer is to clean out the seams and fill with one of the elastic rubber-like compounds now available, eliminating caulking cotton entirely. But these will only hold on *bare wood.* Every vestige of paint and old compound must be cleaned out, and bare wood exposed on both sides of the seam.

This can best be accomplished with your electric handsaw. Use a coarse blade, and set for a depth of cut a little more than half the thickness of the planking.

Now run the saw down the exact center of each seam, slowly and with extreme care. The idea is to take just a little wood off both plank edges. You'll have a tendency to run off a bit to one side, so check as you go, and be sure *both* plank edges are equally clean. The roughened fibers of the wood insure a good bond with the compound.

Depending on the make and the consistency of the com-. pound, it can be applied with a knife or caulking gun. Since the nature of these compounds vary with the manufacturer, *read the directions on the can* before you start. And remember to allow for swelling—don't fill flush with the surface.

IX

Steam Bending and Caulking

I have often noticed that amateur boat builders take fright at the mention of steam bending, in the mistaken belief that it is too involved or too difficult. In fact it generally influences their choice of boat. They invariably choose a V-bottom design over one of round bottom construction, because the latter requires steam bent frames. Most professionals agree that round bottom hulls involve less labor in framing than those with V-bottoms.

Steam bending requires three things—a source of heat, a form of boiler or other container to hold water, and a box or chest into which steam is piped from the boiler. Reduced to its simplest form, this equipment can be improvised from a 5-gallon can over an open fire on the ground, with a short length of garden hose leading from the fill-hole to a simple wooden box a foot or so square and 6 or 8 feet long, which is braced up horizontally over the can.

This rig will serve very well for all the steaming required in building a *small* round bottomed boat, say not over 15 feet in length. But for a larger craft, where the dimensions of the bending stock are much greater, you

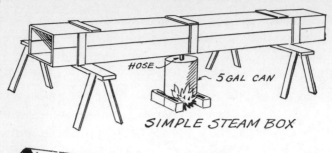

HOSE 5 GAL CAN

SIMPLE STEAM BOX

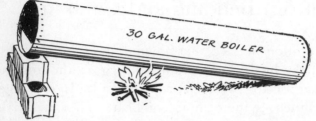

30 GAL. WATER BOILER

MORE EFFICIENT RIG FOR STEAMING
OR BOILING

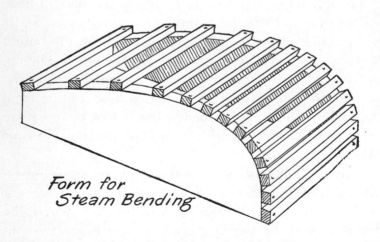

Form for
Steam Bending

150

would naturally require a steam box more on the lines of a professional one, and you'll find instructions for making one in most of the books on boatbuilding.

There is an old rule of thumb to determine the length of time required to make the bending stock pliable enough for working, which is approximately one hour in the steam box for every square inch of cross-section area. This will vary with different woods, of course. Soft soap or rock salt is sometimes added to the water in the boiler to lower the boiling point.

But in the light of the statements I made in the opening paragraph, I am not going to go any deeper into the subject of steam boxes. Rather, I am concerned more with the problems of the amateur who is involved in a project a little less ambitious, the fellow who has but a few frames, planks or other members that can't be bent to the desired curve without steaming. And my thesis is that steaming with a steam box is not an absolute necessity.

Let us suppose you are building a V-bottom hull 14 feet long, and are having difficulty installing the chines. In spite of coaxing, they just can't be sprung to the sharp curve in the forward sections, and you can't bring the ends into the stem for fear of breaking them. What to do? It just doesn't make sense to go to the trouble of making a steam box to bend only two chine pieces.

If the chine lacks but a little more bend to fit, have a helper slowly pour a bucket of boiling water on the chine where the curve is sharpest, while you carefully, and slowly bring the end in to the stem. If the sectional dimension of the chine is small, this will invariably work.

If it doesn't, wrap the bending section with some old burlap, or heavy woolen rags—plenty of them. Then pour on boiling water steadily, several buckets full, and let it stand for 15 minutes. This often does the job where the other treatment won't.

If you are pretty sure that something more effective is going to be necessary, try this: wrap the member *heavily*, with woolen rags, and then with an old piece of carpet, and lay it on the floor. Now pour on several buckets of boiling water, fast, and immediately cover the whole works with a piece of discarded linoleum, oil cloth, or tar paper, or even an old rug. Now let it set for half an hour. The covering keeps the heat from escaping, and drives the steam right into the wood. Work fast when you take the wood out, and bend it into place while it is hot—and you had better wear gloves when you handle it.

I have used this method of bending many times, often when the bend was quite severe, and it was always successful. But where there are many pieces to be bent, and the job becomes more of a production, you need something a little more elaborate. Boiling the wood is just as effective as steaming, and somewhat quicker, so a boiler must be devised.

The best rig I know of is an old 30-gallon household hot-water tank. If you scout around your neighborhood you can generally "promote" one at little or no cost. First you plug up the inlet and outlet holes, and then cut a large hole in one of the ends. Better yet, cut two-thirds of the end away, leaving a shallow lip, as you will see in the illustration.

The water tank is laid with one end on the ground, and

the open end resting on some concrete blocks or bricks, so that it is inclined at an angle of about 10 or 15 degrees. The tank is filled with as much water as it will hold, and a good hot fire is built under it. In place of a wood fire, which must be tended, a plumber's pot heater may be used.

When the water is boiling, and not before, your bending stock is shoved into the tank, as many as it will hold, or as many as you intend using at one session. With some old burlap or rags, close all the spaces around the wood at the tank opening, and let her boil for an hour or so. This method of bending can be used successfully for boats of considerable size.

I know of a 38-foot round bottom auxiliary in which all the frames were bent in this manner. The only difference was that two 30-gallon tanks were welded end-to-end, to take the full length of the timbers.

As I have said before, not all woods are suitable for bending, and some will stand considerable distortion without failure. Stock for bending must be absolutely straight grained, and you should select it with care. Where the bend is severe, great strain is exerted on the frame of the boat, and everything should be well braced to prevent throwing it out of line.

Often it is expedient to bend some member *before* it is fitted into the boat. Here the procedure is to make a form over which the piece is sprung after it comes out of the boiling tank. After it is dry and well-set, it is removed from the form and fitted at leisure.

The exact curve is taken off the boat on a template, by spiling, and transferred to two heavy pieces of scrap lumber. They are then sawn out, and heavy slats are nailed across them, closely spaced, as shown in the illustration. The boiled piece is then bent over the form and clamped in place. Let it set for a day or so, and don't take it off the form until you are ready to install it. The reason is that the wood has a tendency to flatten out when you take it off the form. In fact, it is a good idea to make the form with a slightly sharper curve than is desired, providing it has the same characteristic, so that it may spring back an equal amount.

Caulking

After a hull has been planked up, the bungs cut off, and the plank seams planed smooth, you are faced with the important job of caulking—a job most amateurs approach with more hope than faith in their ability. Actually, caulking is not difficult, it just requires an understanding of the principles and a little serious practice. I would say at the outset that this is one of those skills you can't master by reading a book. To be sure, you can learn the fundamental procedure, but I advise having an experienced man coach you when you start caulking, to correct your mistakes and perfect your technique.

First a word about tools. You'll need a caulking iron and a mallet to drive in the caulking cotton. Irons come in a variety of shapes and sizes, and a professional set consists of 10. Amateurs can generally get along with but one or two, called "Common," "Making," or "Creasing." Thickness of blade determines the size, and you should have a #BB, Crease, 1/32 in. thick, and a #O, Crease,

Caulking Tools
(NOT TO SCALE)

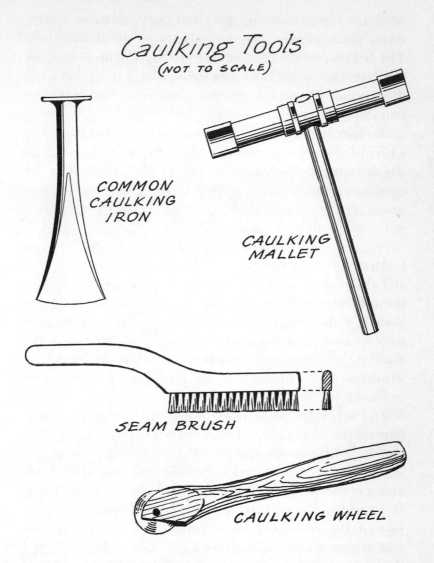

COMMON
CAULKING
IRON

CAULKING
MALLET

SEAM BRUSH

CAULKING WHEEL

1/16 in. thick. You can use an ordinary wooden mallet, but a 12 in. professional yacht caulking mallet is better, and has less "bounce." Boat caulking cotton is sold by the 1-pound bundle, in four grades, and is in the form of a strand. Be sure to get the finest grade, in which the strand is slightly twisted.

Before caulking starts the seams must be prepared, by painting them with a seam brush and thick paint. Caulking starts while the paint is wet, for the paint swells the cotton and binds it, and adds to its life. Some professionals caulk "dry," and paint later, on top of the cotton. I believe in painting before and after. The seam brush is very thin, with short, thin bristles, and really gets the paint deep in the seams.

In the illustration I have shown a cross section of a typical seam. Note that the planks meet tightly, wood to wood, on the inside, that the caulking cotton about half fills the seam, and the caulking compound nearly fills the rest. If the compound were flush with the outside of the planking, it would squeeze out considerably as the wood swelled.

The caulking is first driven in lightly in short bights. The bights should be 1½ in. to 3 in. long, depending on the width of the seam. The wider the seam, the shorter the bights, so that more cotton is driven in.

One of the secrets of learning to be a good caulker is first to learn the correct way to hold the iron. Notice in the illustration that the iron is held between the thumb and second finger of the left hand, while the extended first finger guides the strand of caulking and controls the size of the loops. The hand is *under* the iron, palm up-

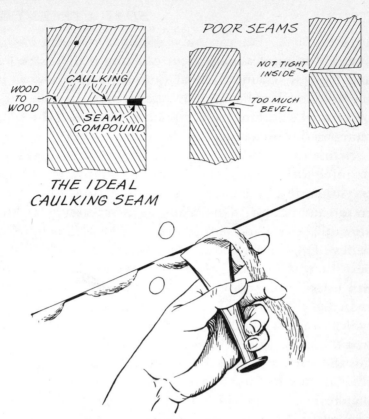

POOR SEAMS

WOOD TO WOOD

CAULKING

SEAM COMPOUND

NOT TIGHT INSIDE

TOO MUCH BEVEL

THE IDEAL
CAULKING SEAM

Correct Way to Hold the Iron and
Tuck the Cotton

Hold Iron Square with
Seam at all times

RIGHT   WRONG

ward, and the strand is always ahead of the iron. Actually, a professional hardly grips the iron at all—it lies loosely in the palm and the fingers merely keep it in position.

When the cotton has been tucked in bights by light blows for a couple of feet, go back and start driving it in. Even though the edge of the iron is slightly curved, hold it square with the work and don't rock it forward or back, or you'll cut the cotton. Drive the cotton home with even, heavy blows if the planking is mahogany, lighter blows if a soft wood such as cedar. The idea is to drive the caulking into a tight bunch, or rope, about halfway into the seam, so that when the wood swells the cotton will actually make an impression, a sort of continuous groove in the sides of the seam that keeps it from coming out.

If the caulking is driven too hard it will go completely through the seam, and if too lightly, it does no good at all. By experience you learn to strike a happy medium. Watch the seam ahead—if it widens more caulking must be forced in, and as it narrows it is thinned out. Try to caulk evenly—the force with which it is driven should be uniform, even though there is more cotton in some places than in others.

The most difficult caulking job to do is in a boat that is poorly planked. A seam that is not tight on the inside, wood to wood, shows that it is a poor job of planking. The cotton can't be driven home, and the only way to prevent such a boat from leaking is to set it in marine glue, a makeshift at best.

Where the planking is less than ⅝ in. thick, the caulking iron is not used. Instead the caulking is done with a caulking wheel, and candle wicking is used instead of the regular cotton. Here the wicking is not in loops, but lies straight along the seam, and it is rolled in by pulling the wheel along, not by pushing, under pressure. In this type of job, the planking seam may be tight outside as well as inside, for the wedge-shaped edge of the caulking wheel can be used to open the seam by running it along *before* putting in the wicking.

As stated before, once the caulking is in, the seams must be painted or "payed" with the seam brush, and then filled with seam compound. There are two kinds of compound, one for deck seams, and the other for hull seams. They are elastic, and stay that way for a long time. Don't use putty—the linseed oil soaks into the wood, leaving the putty hard, dry and inflexible. Fill the seams with compound, using a putty knife, to somewhat below the surface of the planking, to allow for swelling.

X

Wood Finishing

With the completion of new construction or alterations the question of finish arises. Generally speaking, painting is not much of a problem for the amateur of average ability. I have noticed that he generally does a pretty fair job, although he is prone to rely on paint's covering power to hide mistakes and poor workmanship. But bright-work is another matter. Varnish conceals nothing—all the poor joinery, the nicks and dents from tools that slipped, the cross-grain scratches, and the poor preparation of the wood—all are magnified for everyone to see, visible evidence of the worker's skill, or lack of it.

Bright-work requires far more skill, effort and time than painted surfaces, and there are no short cuts that will make the job easier. Most of the work is in preparing the wood for varnishing, and in my opinion, the preparation really begins when the wood is being fitted into the boat.

To explain my meaning, you know in advance just what parts of the structure are to be finished bright, and this should be kept in mind as you work. In the first place, if you are working out of doors, the work should be covered to protect it from the rain. The reason is, that water getting on the bare wood causes stains. On oak, the water makes a black stain that no amount of sanding will remove—it requires a bleach. On Philippine mahogany, water leaves a stain that is often invisible, but which shows up conspicuously after staining and varnishing, and of course it is then too late to do much about it. So if you can't work under cover, see that the wood is protected by a tarpaulin or some sort of cover that will keep off the dew and shed rain.

The next, and most important factor in protecting wood that is to be varnished lies in the way you work, and the way you handle your tools, and herein lies the difference between amateur work and professional. Just remember that every mar, scratch, dent and nick that occurs while you are working, will eventually have to be taken out by an awful lot of sanding, if at all. Constantly guard the face of the wood at all times—don't let anything strike it, and don't drop it carelessly, but lay it down face up. If you have to strike it with a hammer to force it into place, hold a clean block of wood on the face to take the blows. *Never* put a C-clamp on *any* member, whether it is to be painted or varnished, without first putting a block under the clamp jaws. Clamp marks cannot be removed short of planing them out, and this is an occurrence invariably found in amateur construction.

Another source of damage to the face of the wood is the improper method of driving nails. *Never* drive a nail completely home with the hammer alone—the face of the hammer will always make disfiguring dents with the last few blows. The proper, professional way is to drive the nail only until the head lacks ⅛ in. or so of being flush with the face of the wood, then use your nailset to finish driving it home.

Hold the nailset correctly—between the thumb and the first two fingers, with the third finger holding the end on the nailhead, and the little finger resting on the wood for steadying.

The first step in the actual preparation of the wood for finishing is the sanding—a tedious, irksome job which cannot be slighted if you are to achieve a professional finish. There are a number of types of sandpaper available with various kinds of abrasives. The old fashioned flint paper is a soft abrasive, which fills up rapidly and lasts a very short time. "Production" paper, whose abrasive is aluminum oxide, is most commonly preferred because it is "sharp," cuts much faster, and lasts considerably longer.

It is a mistake to use a coarse sandpaper on bare wood, in the belief that it cuts faster and therefore shortens the job. Coarse paper is strictly for such jobs as cutting down heavily painted planking, etc. The first sanding should be done with a *medium* grit, followed by fine grit and finishing with very fine.

Almost everyone has heard it said that you should "never sand across the grain." But the trouble is that many amateurs pay no heed, believing it to be unimportant. Actually it is mighty important, as any good cabinetmaker will tell you. Just one swipe across the grain will leave a scratch that you may not be able to see. But after the wood has been filled, stained and varnished, in spite of much corrective sanding with the grain, it will stand out conspicuously, an unsightly symbol of carelessness. So take this old maxim seriously, as a careful worker should, and don't ever forget it.

Not all woods are suitable for varnishing. Bright work is generally reserved for those woods whose grain and color have an inherent beauty worth preserving. Mahogany, red cedar, cherry, walnut, and in specific cases, spruce, are most often finished bright. Oak is often varnished, but maintenance is a problem. Once the varnish is abraded and water gets to the oak, unsightly black stains appear. All varnished surfaces require constant attention while the boat is in use, but this is more true of oak than of any other wood.

Oak, cedar and spruce require no preliminary treatment for varnishing other than sanding. Mahogany, however, requires a filler, and invariably a stain, and it takes good judgment and experience to achieve a professional finish.

Some of the *true* mahoganies are best finished with varnish alone, while others are commonly filled and stained. The mahoganies most commonly used in boats are the Honduras, and the Philippine mahogany, which, as I stated earlier, is not a true mahogany. Honduras mahogany is stained in order to obtain a uniform color, since one piece may have a different color from another. Philippine mahogany has a very poor color to begin with, which is improved but little by varnish alone. Its only resemblance to true mahogany lies in its grain, therefore it requires staining to give it the mahogany color we are accustomed to.

Philippine mahogany has a soft, open grain, so it requires the use of a wood filler to close the pores and provide a good base for the varnish. Fortunately, all the marine paint manufacturers make a paste wood filler in which the stain is incorporated, so you can fill and stain in one operation. All of them give you your choice of color, light mahogany or dark. Some makers label the "light" as "red."

The dark mahogany is more of a brown than anything else, and is not a very pleasing color to begin with. Since all mahogany darkens with age and exposure to sunlight, the dark mahogany stain is generally used in staining new wood to match some that is old. The red mahogany is most commonly used in the average boatyard today and gives the brilliant, high color that is found in all new construction.

The combination wood filler and stain is in the form of a thick paste. Take a small portion from the can and mix with turpentine in a separate container to the consistency of cream. The paste takes considerable stirring to dissolve completely, and it is better to make a thin, paint-like mixture than a heavy one. It is applied to the wood with a brush, allowed to set and strike into the wood, and is then wiped off with a coarse rag.

The trick here is to wipe the mixture completely off the surface, leaving it only in the pores and grain of the wood. Treat only a small area at a time, and wipe it off the moment the stain-filler "flats," or loses its shine, as it strikes into the wood. If you wait too long, you'll have difficulty wiping it off. Always rub *across* the grain, so that the pores will take up the filler, and use lots of elbow grease. If the mixture is too thick, and if you don't wipe it off sufficiently, too much pigment will be left on the surface, the natural grain of the wood will be hidden, and it will have an artificial, painted look. So don't lay the stuff on in puddles, wipe it off promptly, and have plenty of clean, coarse rags. An old woolen suit makes excellent rags for the purpose.

Now if you read the directions on the can of filler-stain, you'll notice that they say to wait 24 hours before varnishing. Here's where you can run into trouble. Some makes of filler-stain use an oil vehicle that is very slow drying. I recently saw a new motor-sailor on which all the exterior varnish peeled off in large sheets within a matter of weeks after completion, and the builder had to refinish the entire job, at his own expense of course. The varnish peeled for one reason only—the filler-stain was not completely dry, or "cured," Since this was not an isolated case, I earnestly recommend that the stain and filled wood be allowed to set at least three days before varnishing—even longer to be on the safe side.

The filler and stain raises the grain of the wood and leaves the surface somewhat fuzzy. So before varnishing, and after it has cured, it should be given a light sanding with a very fine-grit paper. Finally, go over the whole area with a soft rag dampened with turpentine to remove every vestige of dust.

When it comes to varnish, you have many makes and kinds to choose from, and I wouldn't recommend one over another. Just be sure the spar varnish you pick is designed for marine use, and don't buy it on the basis of low cost. The best is none too good for a fine boat.

With the filler-stain well cured, the first coat of varnish should be thinned slightly with *not more than 10%* of turpentine. Too much turpentine destroys the bonding power of the varnish, and it will sag badly. All subsequent coats should be straight varnish.

Varnishing is an art that one acquires slowly, and with patience. You can't learn it in a day, for there are several factors involved which require coordination. First, the weather, that you can't control. Warmth and low humidity can be had indoors, but outside you take it when and if you get it. Varnishing out of doors can't start until the morning sun has dried up every vestige of dew, and it must cease at least two hours before the evening dew falls —which is a lot earlier than sundown, generally.

Secondly, you must have good brushes, designed expressly for varnishing, and kept immaculately clean. Finally, you have to know how to put the varnish on. The inexperienced do one of two things. Having heard that it should be "flowed on," they load the brush excessively, take the word "flow" literally, and fail to brush out the already-too-heavy coat, with the result that it sags, runs, and lays in pools, dries unevenly, and with wrinkles. Or having observed this, they go to the other extreme and "scrub" with the brush, starved for varnish.

You eventually learn to strike the happy medium, to put on a thin film of varnish *evenly,* with a minimum of brushing, so that it does not sag or run, and sets evenly. Varnish dries from the surface inward, and too heavy a coat will skin over on the surface before the under part has set. That is what causes wrinkles and sticky beads that never harden. Don't dip your brush deeply, and wipe the surplus out on the edge of your can. *Never shake or stir a can of varnish,* or use the brush vigorously —that causes bubbles which are hard to eradicate.

The standard specification for a fine varnish finish calls for five coats, with a light sanding with very fine paper between coats. Remember, nothing ruins a good job like sanding before the varnish has thoroughly hardened, or failing to wipe all the dust off before varnishing. Read the instructions on the can before you start varnishing—they weren't put there because the makers think you are an ignorant tyro, but because all varnishes differ in their composition, and each requires different handling.

One final tip in achieving a fine finish. Often in interior cabinetwork a high gloss is not desired, and sandpaper would be too harsh, no matter how fine. In a good paint store you can get a small quantity of powdered pumice and a can of lemon oil. Moisten a soft cloth with the oil, dip it lightly in the pumice and *gently* stroke the varnished surface, with the grain. Then wipe it off with a clean, soft cloth. The result is a satin finish that glows instead of glittering, greatly enhances the grain of the wood, and looks like a professionally-finished piece of fine furniture.

XI

Gold-leaf Application

In the modern trend of "do-it-yourself," more and more boatowners are doing their own overhauling and maintenance work—or as much of it as the boatyard rules will allow. But when it comes to lettering the boat's name and home port on the transom the conscientious owner will call in the local sign-painter for a really professional job, especially if it involves the use of gold-leaf. It is true that many an amateur can work over or repaint a *painted* name that was originally put on by a professional, but gold-leafing is a highly specialized skill invariably beyond the limits of his experience.

Nevertheless it is entirely feasible for the amateur to turn out a fair job of gold-leafing, provided he understands the fundamental techniques, has the proper equipment, and is reasonably dexterous. Two things should be firmly planted in his mind right at the start—he is at best strictly an amateur, and his time isn't worth a red cent!

Gold-leaf is supplied in "books" of 25 sheets, each leaf being 3⅜ by 3⅜ inches, and the price of a book is currently under $2. The 22K gold is beaten to a gossamer thinnesss that is really amazing. The slightest movement of air, even your own breathing, will send it flying away. For this reason it takes considerable skill and experience to successfully transfer the loose leaf from the book to the surface to be embellished.

Fortunately for the amateur, however, gold-leaf is also supplied in a form known as "Patent," in which the leaves of the book are tissue paper lightly coated with wax. The gold-leaf adheres lightly to the waxed tissue and is thus prevented from blowing away. To apply it you cut off a piece of the gold-surfaced tissue with scissors or razor blade, lay it face down on the surface and then peel off the tissue.

The exact area to be gold-leafed must first be covered with a "size," to which, when it reaches a certain state of "tackiness," the gold will adhere. There are two kinds of gold size, each having specific characteristics and uses. "Slow" size is a rather thick, pigmented medium (generally a vivid yellow), somewhat in the nature of a heavy enamel. It is extremely slow setting, requiring two to six days to reach the proper tackiness for gilding. It is used primarily on carvings and decorations rather than for lettering, or where there is a very large area to be covered. Because of its deep gold color, any cracks, scratches or bare spots in the gold-leafing job are not very noticeable.

Much faster and easier to work with, Japan Gold Size is recommended for the amateur. This is a thin, transparent varnish-like liquid which sets up to its tacky state in a matter of 15 minutes or so, depending on the temperature and atmospheric conditions. Because of the speed with which it dries only one letter should be sized at a time.

Hastings' Patent Gold-leaf and Japan Gold Size may be obtained from most of the better paint stores or dealers in art supplies. From the latter you should also obtain a couple of red sable or camel hair lettering brushes, size Nos. 4 and 6. These are needed not only to apply the size, but to pick up bits of loose gold-leaf for covering holidays and the inevitable breaks that are bound to occur in the gilding. With the addition of a can of talcum powder and a small package of absorbent cotton your list of supplies is complete.

Let us assume that the transom of your boat is varnished mahogany, and a number of years ago the name was put on with gold letters outlined in black, the standard practice. From the yearly sandpapering and revarnishing the lettering has become dingy, scratched and unsightly, and the time has now come for a complete refinishing. Sand the whole transom with very fine sandpaper to your satisfaction and revarnish as usual, being particularly careful to see that the area covered by the lettering is absolutely slick and smooth. Pick as near perfect a day for gold-leafing as possible—dry and windless. Rig a tarpaulin if necessary to serve as a windbreak and to shade the working area from the sun.

With some talcum powder on a piece of absorbent cotton rub lightly the entire area covered by the lettering. The purpose of this is to prevent the gold from adhering *outside* the letter. With your lettering brush paint the first letter with the Japan size. Don't load your brush— the size is so thin it will run very easily on a vertical surface. If you run over the edge a bit it isn't too important, for the black outline will probably hide the mistake. If you really make a mess of it, wipe it off with turpentine, then re-powder the area and try it again.

Now comes your first major problem—deciding when the size has reached the proper tacky state to receive the gold-leaf, and you'll have to experiment. I might say it will take 10 to 15 minutes—and I could be wrong. It all depends on the temperature, humidity and the thinness of the size. Here's a way to test it—touch the letter lightly with your knuckle. You should hear a slight "snap," and if any size adheres to the knuckle it is too soon. If you applied the gold too soon it would slide on the size and break or crack. If you wait too long the size might be too dry to properly hold the gold. The natural tendency is to do it too soon.

Referring to the illustration, let us assume that the first letter is "E." Cut a strip from a leaf of your book a little larger than the vertical leg of the letter. Starting at the top, lay the strip on in a sort of rolling motion. Note that the second finger positions the strip and serves as an anchor. Do it lightly—any real pressure will cause the strip to move, and cracks will appear in the gold. The gold merely has to touch the size to adhere. Now stroke the strip lightly with a tiny ball of cotton, being careful not to touch the rest of the sized letter. Peel off the strip of tissue. If any holidays or cracks appear in the gold don't worry—they can be touched up later.

Now cut a strip for the top horizontal and apply it in a similar manner, starting at the *right* and laying it on to the left. Remember you always cut your gold somewhat larger than the area to be covered, to give yourself plenty of leeway in positioning it. Never try to lay on an entire sheet at once.

When the entire letter has been covered, pick up a stray bit of gold with the tip of your other brush (the

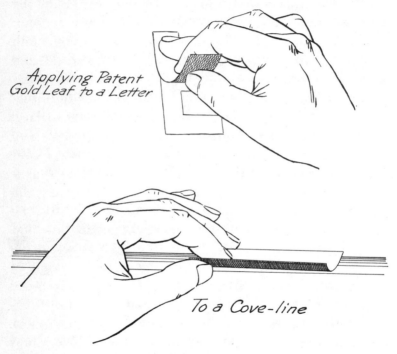

Applying Patent Gold Leaf to a Letter

To a Cove-line

clean one), and lay it on any crack or holiday that may appear. If there are now no breaks in the gold anywhere take a small wad of the absorbent cotton and stroke the letter lightly, outward in all directions, to smooth it down and to remove the free gold from the edges. Now size the next letter and proceed as before.

When the last letter has been gilded all that remains

is to outline the letters in black. Just any old black paint will not do—use only Japan color which comes in a tube and can be had at any paint store.

Names and registration numbers are not the only places where gold-leaf is required. Many boats, both sail and power, have a hollowed-out cove-line just below the sheer, dying out just short of the stem and the transom. Often it terminates in an arrowhead and feathered tail. Originally these cove-lines were gold-leafed, but all too often you'll find them painted over and I suspect it was due to the owner's unfamiliarity with the art of gilding.

Although here you are dealing with a curved surface, the procedure is much the same as with lettering. Size not more than 2 feet at a time, and cut your srtip of gold about 1/4 inch wider than the cove. Hold the strip as if you were about to roll a cigarette and lay it in from the top edge downward.

Some boats have carved trailboards or other bow and stern ornamentation. The technique is still the same— size and gild a small portion at a time, each section of gold-leaf overlapping the one previously applied. Since the gold is 22K pure, in theory it needs no finishing. However it is standard practice to give it a coat of clear spar varnish where it is to be exposed to the weather. Although it tends to darken the color of the gold, it protects it from abrasion and wear.

The professional sign painter and the picture-frame maker often bring the gold up to a very high gloss by rubbing it with a burnisher—a rounded agate tool. However this is a chapter we might as well skip, for we are still, remember, strictly amateurs!

Index